Praise for Brian Keith Jackson's
THE VIEW FROM HERE

Winner of the American Library Association Literary Award for First Fiction from the Black Caucus of America

"A tender tale of forgiveness . . . Jackson has a good ear for dialogue, lovingly capturing the rhythms of Southern African-American speech."   —*The Washington Post Book World*

"Vivid . . . The author Jackson brings to mind most readily is Alice Walker. . . ."   —*Los Angeles Times*

"Mr. Jackson's prose has a visceral pungency. . . ."
    —*The New York Times Book Review*

"Jackson is a storyteller; he has the rhythm of conversation and the ability to take a simple story, build tension, reveal truths. . . ."   —*Denver Post*

"A stunning literary debut with a cast of characters whose roots are as deep as the landscape they inhabit. . . . The author's gift for dialogue . . . brings these characters to life. . . . A refreshing voice."   —*Bay Area Reporter*

"The dialogue and dialect are right on target."   —*USA Today*

"An extraordinary debut . . . A formidable craftsman and exceptionally gifted storyteller. [Jackson] has written a haunting story."   —*Publishers Weekly* (starred review)

"A beautifully told tale. Highly recommended." —*Library Journal*

"Heartfelt about the strength of poor black women and the weakness of the men who oppress them . . . More touching than Alice Walker's novels on the same theme."—*Baltimore Sun*

Books by Brian Keith Jackson

Walking Through Mirrors
The View from Here

Published by WASHINGTON SQUARE PRESS

For orders other than by individual consumers, Pocket Books grants a discount on the purchase of **10 or more** copies of single titles for special markets or premium use. For further details, please write to the Vice President of Special Markets, Pocket Books, 1230 Avenue of the Americas, 9th Floor, New York, NY 10020-1586.

For information on how individual consumers can place orders, please write to Mail Order Department, Simon & Schuster Inc., 100 Front Street, Riverside, NJ 08075.

Walking Through Mirrors

Brian
Keith
Jackson

WASHINGTON SQUARE PRESS
PUBLISHED BY POCKET BOOKS

New York   London   Toronto   Sydney   Singapore

This book is a work of fiction. Names, characters, places and incidents are products of the author's imagination or are used fictitiously. Any resemblance to actual events or locales or persons, living or dead, is entirely coincidental.

Originally published in hardcover in 1998 by Pocket Books

WSP
n A Washington Square Press Publication of
POCKET BOOKS, a division of Simon & Schuster Inc.
1230 Avenue of the Americas, New York, NY 10020

ISBN 978-0-671-56894-8

First Washington Square Press trade paperback printing August 1999

10 9 8 7 6 5 4 3 2

WASHINGTON SQUARE PRESS and colophon are
registered trademarks of Simon & Schuster Inc.

Cover design by Jeanne M. Lee
Front cover photos: top, © Bob Sacha/Aurora/PNI; bottom, Joanne Dugan/ Graphistock

Printed in the U.S.A.

Acknowledgments

No journey is traveled alone. That being so, I wish to extend my gratitude to:

Calvin Baker; David Chestnut; Phillip Christian; Paul Cooper; my editor, Greer Kessel, Amanda Ayers; and the Pocket staff; Katherine O'Moore-Klopf; Juan Gaddis; Janet Hill; my parents, A. S. Jackson III and Dr. B. Rene Jackson; Lloyd S. Jolibois Jr.; Jeanne M. Lee; Bruce Morrow; David Paul; Teachers & Writers Collaborative; Carl Swanson; my agent, Emma Sweeney/Harold Ober & Associates; Theresa Zoro; and, as always, my entire family.

Shipley's Donuts of Monroe, Rue del la Course of New Orleans, and Drip of Manhattan, for allowing me to sit with my thoughts as long as I pleased.

Please be patient with me,
God is not through with me yet.

—GOSPEL HYMN

He was returning home, yet he had always wondered exactly what that was. "Home is home is home," he remembered Charles saying. For Jeremy, home was where the largest amount of memories existed, a place you're free to flee, and he did it well. Now he was going back there, but would going back mean moving forward?

"You must plow your own rows, mindful of pebbles 'long the way." That's what he heard Mama B saying as he stood behind the rocker in his filled yet barren apartment. In times such as these, it was her spirit that kept him going. Now and always, he plowed on.

A deep breath.

Another. Then he held it as if it were the life it possessed.

He released the rocker, and the sound of its sway filled his ears, drowning out the faint hum of electricity. He took one long look around, extending his gaze beyond the open space in which he stood to the other rooms, confined by either drywall or exposed brick. He could see every one of the rooms without being in them, as though looking at blueprints and filling in the thin blue outlines with the things he'd acquired over the years. The gaze was that of someone seeing something for the first—or last—time. He again looked at the rocker, which had stopped moving and now stood stalwart.

Then he began the journey.

When the elevator opened, the doorman came to get his bag.

"I've got it, Jimmy."

"Off on another trip, Mr. B?"

"Yep."

"Well, have a good one."

"I'll try."

Whenever he left the building, he could depend on some exchange with Jimmy, always ending with his saying "I'll try." Some things never change. Today it was pleasing to have some sense of certainty.

The driver put the luggage into the trunk and Jeremy settled into the backseat. He hadn't slept and his bloated face had yet to regain its normally firm shape. His sunglasses, centered on his face, became another reflection, providing a sepia view.

"Would ya like a paper or magazine? I've got the *Times* or the *Post*," said the driver, attempting to scope out what sort of passenger Jeremy would be. "I always say, 'tween the two, you can kinda get an idea of what's goin' on. One makes ya feel smart; the other makes you glad ya are. 'Course, I leave it up to you to decide which'n does what."

Still no answer from the backseat.

"D'ya mind music?"

Still no answer.

The driver had seen many like this and had often found himself thinking, *So dat's how it's gonna be, Mr. Big Shot? Mr. Just-Get-Me-Where-I'm-Going-and-Don't-Talk-to-Me?*

But those thoughts were inaccurate on this journey. Jeremy wasn't that sort of passenger at all. Yes, he took car services and had his moods, but ignoring wasn't something he practiced—often.

"Do you have a paper I can borrow?" asked Jeremy.

"Yes, sir."

"Please, don't call me sir." The driver could have easily been his father, in age if nothing more.

"Whatevah ya say," said the driver, glancing back at Jeremy. "I offered ya one earlier. Used some of my best material, too."

"Sorry. My mind was elsewhere."

"Really? Where's that?"

"Far away, but never far away."

The driver glanced in the rearview. He passed the paper and watched as Jeremy hid behind the news of the day, noting that this was probably best, a silent fare.

For as long as he could remember, he had been told that his eyes were those of an old person. When he was born, they looked out on the world as though they had seen it all before, and the haze that coated them wasn't at all one of youth or confusion.

5

He was an old soul and set in his ways from the moment the warmth of the womb changed to the slap of room temperature.

Having always been old, he realized that the things we love the most are things we've had the longest and the things we've longed for but never had. Love lost and love desired carry the same weight, for no price can be placed on time spent.

Jeremy was no stranger to loss.

He was left with Mrs. Bishop, his father's mother. He didn't realize she was his grandmother; titles matter little where nurturing is concerned. Mrs. Bishop had three children, yet only Jess remained with her while Jeremy was there. She was the oldest and stayed near home, while her brother and sister went off to conquer the world.

Mrs. Bishop and Jess were like pods bursting to free their seeds, providing life for the next. But it was life that showed Jeremy that often birds had other plans, nabbing the seeds for nourishment of a different kind. He and he alone called Mrs. Bishop Mama B. *Bishop* never could find its way off his "lazy tongue." Time strengthened his tongue, but the taste for the name remained.

On numerous occasions, he'd heard that his first word was *mine*, but because of his lazy tongue, they believed him to say *Mama*. Though most women would coo at the thought of a child's first word being *Mama*, he was certain no joy was found in what they believed to be his first proclamation, for it was nothing more than a constant reminder, another wound to bleeding hearts. They soon realized that he was saying *mine*. He would grab his Mama B or Aunt Jess and say, "Mine."

Jeremy's mother died a few hours after he was born. One soul for another. She was holding him when she died. No one was present. He was the last person to feel her alive, yet too unaware to appreciate it.

He wondered if she smiled when she held him. He wondered if she thought he was the most beautiful creation in the world. He wondered if she felt the angel in her arms in the same way that she felt the one hovering over the hospital bed. Did she gently gloss his cheek with the smooth tips of her fingers as he'd seen other mothers do with their newborns? He could write a list of wonders that never wandered far from mind.

Wondering and wandering were his most familiar friends.

It was Jess that found him in her arms, his tiny hands and mouth grasping for a nipple numbed, nurturing soured. He was the last to fill that heart that had kept him alive, and though his eyes had not opened and he couldn't rightly say he remembered their time together, he could say that that moment was one of his prized possessions. Still, it was nothing he could or would ever hold.

Mama B called me Patience. I knew my name was Jeremy, but she called me Patience.

Many, on first hearing it, would question, "Patience? Why you callin' that boy that? That's a woman's name."

And I suppose they were right. That name does suit a woman. Yet, like a child too young and unaware to realize he is poor, I didn't know that I wasn't a woman. Sure, I knew they were different, but I didn't know the difference. Ignorance is bliss.

Sometimes.

"Mama B?" I asked one evening. "Why do you call me Patience?"

"Why?" she asked, stopping the sway of her rocking chair. "I been callin' you that all ya life. Don't you care for the name?"

"It's fine. It's just that my name is Jeremy."

I rarely questioned her about her actions. She was an elder, but when you're given a girl's name, time warrants a defense when prosecuted by unruly peers.

"See, Jeremy is ya given name," she said, relaxing into her rocker. "But through the course of life, people always take on other names, and then you grow into it, just like you call me Mama B. It's just more personal, a show of affection. When

I was a chile, I was called Sister 'cause I had all brothers and it was easy for them to call me that 'cause I knew who they meant. Now, anybody that knows me from them days still calls me Sister. It's something familiar, and people like that."

"Then how did you come up with Patience?"

"Well, it just seemed to suit you. Patience, you see, comes from being able to wait on things and being still. You ain't never minded being on ya own. You were a still chile—still are.

"You showed patience right from the get-go. When you were just a wee little something, I had to check on you to make sure the crib hadn't gotten you. I'd put my pinky finger under ya nose, just to be sure, and them quiet l'il breaths were just as faint, but they were breaths nonetheless.

"Lawd knows how I prayed that nothing would happen to you, 'cause I'd never be able to live it down. But you'd just sleep, all sprawled out like you owned the world."

The telling of it seemed to take her back to a simple time, a time with no questions, just reassurance. As the recollections passed before her, a smile filled her lips. It seemed like a fond memory for her, though it sounded like it was a burden at the time. Perhaps that is what retrospect is once it truly becomes that.

Retrospect.

"Oh, yes. See, most babies—least any I've been 'round—will cry when somethin's wrong. But not you. It was like nothin' was ever the matter with you. Just too good to cry. Most babies will test their lungs when they get hungry in the night, but not you," she said, striking the match against the side of the space heater and lighting her cigarette. She took a long drag, and as the smoke slid out of her mouth, so did the words. "It was like you knew the bottle was comin'. But it would scare the dickens out of me, 'cause the bottle was all you had and I had to see that you were eatin'. I ended up havin' to just put the bottle in the crib with you.

"This one night I got up, worry had me tossin', and I looked *in on you. I pulled my rocker in that very room there, where your daddy used to sleep, and looked at you for the longest while. You never rolled over, just stayed still, stretched on your backside. When you finished one bottle, you'd throw it to the side. And I don't mean just drop it—you'd throw it."* She laughed until the laughs became coughs. She reached down in her housedress pocket to pull out the inhaler, then shot the contained breaths into her mouth until they supported her lungs, returning the smile to her face.

"Yes, you'd throw it, and without as much as a peep or the openin' of an eye, you'd feel around until you found the next one, then the next. Three bottles a night in that crib, but never did I see you reach for an empty one. Not one time. You've stretched out now, but you were a big baby. Had more creases than a skirt."

When she finished the explanation for my name, she finally took notice that I was in my pajamas—my PJs. They were the kind that encased the feet. They shielded mine from the uninvited drafts that even the finest old houses take on.

"You all situated for bed?" I nodded my head, indicating that I was, still pondering the story of my name. *"Did you wash out your ears?"* Again I nodded, with a prelude smile on my face. *"And behind them? Let me see."* And she would grab me and look at my ears, always saying, *"So clean I can see ya thinkin' muscle, and it looks to be workin' overtime."*

This had become a nightly ritual with us, and we played our roles to perfection. But her story had made me lose my center, and the words planted me next to her rocker, unable to uproot.

"Why the long face?" she asked. *"You look like you just lost your best friend."*

"You're my best friend."

"And you mine, Patience. But you haven't lost me."

I didn't say anything to that. I just stood there, holding the arm of the rocker, displaying that stillness that had garnered me my name.

"What's on your mind?" she asked as she pulled me around to the front of the rocker. "Patience?"

"Ma'am?"

"Speak up, now. I said, 'What's on your mind?' "

"Nothing."

"Now don't stand there lookin' like that and tell me 'nothin'. Somethin' must be toilin' up there; ya face is longer than a Easter sermon."

I stood looking at her. I considered making up some question or smiling as though I had just been longing for some more attention, but I couldn't bring myself to do so.

"Is my mother ever gonna come see me?" I said, waiting for a story to present a fond memory in her eyes.

"No, sweetheart. She can't come see you."

"Doesn't she want to see what I look like?"

"Your mama can't come see you."

I knew I should leave it at that, let the worry rest. I could see that Mama B wanted me to leave it at that, and it hurt me to hurt her by continuing, but . . .

"Why?"

"Now, you know ya mama's up in heaven. See, she felt that she could keep a better watch on you up there. Up there with God."

"Will I go up to heaven soon and see her?"

"When ya time comes, you will, and then you can see her."

"You think she'll be happy to see me?"

"Oh, Patience, all of heaven's gonna be happy to see you."

Mama B's voice began to quiver like the last leaf of fall, but this time it wasn't from attempting to stifle a cough. Before it broke completely, she told me that it was past my bedtime and that I should get my rest.

I left her with a kiss and the sound of my second feet sliding against the grain, a noise that could tighten a sagging jaw.

She lit another cigarette and began to cough as she pulled her Bible off the nightstand. The Bible was almost a foot thick. Though I strained whenever I tried to lift it, it seemed to fit comfortably between her legs, filling the fall of her housedress.

I pulled back the generation of quilts that covered the bed and climbed underneath them. They weighed heavily on me, so much so that I could barely move from side to side—not that I wanted to. No, I wanted the weight of warmth all over me, capturing me like the sight of a star in a cloudbreak sure to come.

"You need some more blankets?" she asked.

"No, ma'am."

"Now, I can pull some out from under the mattress if you need them. It's no trouble."

I told her that I had enough, and she turned back to her reading. But periodically she kept looking in, if not at me, then in my direction. I could see her, but she couldn't see me. She was in the light, I in the dark.

"You need the heater turned up?" she asked, looking yet again.

"No, ma'am. I'm fine."

"Now, don't forget to say ya prayers."

"I won't."

"Awright, then. Have a good night's rest, and I'll see you in the mornin'," and after a heartbeat, "God willing."

I lay there for a while, looking at the space heater in my room, with its blues and oranges and yellows, different, yet all shooting from the same sprout of flame. I could hear the heat's scurrying murmur. It sounded like wind would sound if you could hear wind. But wind is nothing more than a feeling; the sound is made by the things it passes through.

"Mama B?" I said, almost fearful of disturbing a moment of solace.

"Yes, Patience?"

11

"Do you think God was sleeping?"

"What's that?" she asked, seemingly annoyed by the asking.

"When my mother died?"

"No, sweetheart. He wasn't sleepin'. God never sleeps. He's always watchin'."

"That's what I thought. Good night."

" 'Night."

I rolled over like a wave under the blankets and faced the wall. The wood paneling seemed to be breathing and the manufactured knots on it curved before my eyes as I made every attempt to stare through its plies. I could feel Mama B's eyes transfixed on me, for by now they had adjusted to the darkness. I heard the big Book close and the rocker wince. She walked into the room and made certain that the quilts covered me. I kept my eyes closed when she did this. Though I knew she couldn't possibly believe me asleep, she said nothing.

As she walked away, I picked up my head, straining my neck like a turtle coming out of its shell of protective burden. She turned up the heater and her face glistened in the glow of iridescent warmth. She slowly returned to upright and placed her hand on her hip. She stood there for a moment before making her way back to the rocker, back to the Book, back to moving her lips. But her eyes weren't on the ancient words; they were closed to a crow's-feet squint. I knew she was praying, and I was certain it had something to do with me.

Something to do with patience.

I put my head back on the pillow and again focused on the wall. After a while, I too closed my eyes, but I didn't say my prayers. I didn't forget. I just didn't say them. If God was indeed always watching, always awake, then He would know that I hadn't said them.

I wanted Him to know.

chapter 2

∽ ⌒

The flight arrived in New Orleans, where it would connnect with an express commuter. Once Jeremy boarded, it appeared that he would be the only passenger, allowing him eighteen other seats of peace. But the pilot seemed to know him.

"J.B.?" That caught Jeremy's attention. He associated that with his school days.

"Yes?" said Jeremy with that casual reserve that always sprang to attention when he heard his name yet couldn't summon recognition of the face that voiced it.

"Boy, is that you? Carl Landry—we went to BH together."

"Right. How's it going?"

Jeremy removed his sunglasses so he could have a better look, hoping to place this face. Though he couldn't, his voice and manner never gave a clue. He had steady eyes.

"Hey, I can't complain—and who wants to hear it? Grin and bear it, that's what I say; grin and bear it."

"I suppose you're right," Jeremy chuckled, pleased to hear someone else articulate the sentiment.

"Things are swell, though. I'm a pilot now," said Carl, gesturing to the wings piercing his jacket near his heart.

"Well, I should hope so, or we're in a awful lotta trouble."

Jeremy was surprised how quickly he'd slid into Southern vernacular. It usually presented itself only after several drinks. He'd been told numerous times how he didn't have a Southern accent.

He would always respond with a look that, given its angle, was either a smirk or a smile. It was as though Northerners believed twang hung over the South like moss and lichen covered trees.

No one he knew had a twang. Yes, the words might slowly slide out of mouths like molasses, but that was due more to demeanor than to diction. It wasn't so much that the endings of the words were chopped off; it was that unfamiliar ears failed to realize that all endings were tucked deep in one's chest.

"I'm always hearing great things about you. My mom keeps me posted. She's always flipping through some magazine. You're doing pretty good for yourself up there, hanging out with celebrities and all."

"No, I can't complain either," said Jeremy. He had heard this sort of comment more often than he cared to. Yes, he had become somewhat successful behind the lens. But he knew the largest fame of all was hometown fame, and home was where he was heading.

"I'm living down here in New Orleans now. Got a nice place in the Garden District, Second and Magazine. Right around from Anne Rice. I don't read her books or anything like that, but she helps the property value. Still, I can't seem to get away from home. I keep a little place there. Fly back and forth and back and forth. Coming and going. I run into a few people from school down there. Carla Samms is married, studying at Tulane. She married Blake Bradley. He's with a law firm, making good money—which if you ask me, *good money* and *lawyer* shouldn't be mentioned in the same sentence. Say, if you ever wanna come down for a visit, just give a ring. We're listed. Mardi Gras is a bit of a mess, but maybe Jazz Fest. We've got plenty of room. Hotels these days'll make you break the bank. Not like you can't afford it or anything."

Jeremy just sat, his back pressed tightly to the seat. Carl had his arm on the seat in front of Jeremy and leaned over, trapping him there.

"What brings you back down here? You here on a shoot?"

"Uh, well, Carl . . . I'm here for my father's funeral."

"Jesus Christ!"

"Not hardly," said Jeremy. "More like St. Christopher."

"No. I mean . . . I'm sorry."

"You didn't kill him, did you?"

"What?" asked Carl, confusion coating his face until Jeremy's eyes let him know that he was attempting a joke. A courtesy laugh bounced through the space and though they were in the plane, pressure of a different sort filled the cabin. "Awrighty, then. Sorry about babbling on like that. You probably just wanna be alone with your thoughts. You know all the rules, so I don't guess I need to tell 'em to you. Just don't tell anybody. I'll get the axe."

"Mum's the word."

"Sorry, we don't serve drinks on these commuters."

"That's fine." Jeremy laughed. "I'm sure I can wait."

"Awrighty, then. Just sit back and enjoy the ride. It should be a fairly smooth flight in."

"It's in your hands."

"That's what people like to think, but it's in God's hands." Carl raised his eyebrows and rolled his lips inward to emphasize his statement. Then he ducked into the cockpit and slid the doors closed behind him.

Jeremy buckled his seat belt and again centered his sunglasses on his face. The sky became less blue as he mumbled, "Touché."

The plane slowly began to taxi. He watched the blur of heat distort the runway, his last escape before heading on a journey from which he didn't quite know what to expect, yet fully aware that perhaps the unexpected was what a journey truly entailed.

The commuter flight was nothing more than a grasshopper's leap; a quick up-and-down affair. Within forty minutes, the geometric pastures of cotton and rice gradually gave way to the speck of city.

The airport runway consisted of a U-shaped strip of asphalt next to Chesterfield Park and the golf course. Seeing the pristine shorn fairways and greens reminded him of how he knew Carl.

Carl had been the youngest golfer in course history to sink a hole in one, winning a car for his efforts. As if that wasn't enough, he donated the car to the charity sponsoring the tournament. Jeremy didn't see the hole in one. He and Paul hadn't left the beer tent—par for the course.

The airport had a restaurant downstairs and a bar upstairs, but it was a small building and the term *concourse* didn't apply. It was at this very airport that he had parked his car on many an occasion to watch the planes take off. He loved to watch them take off, but never once did he care to see them land. Now, he was landing.

When the plane came to a complete stop, the first thing that caught his eye was the sign that read WELCOME TO ELSEWHERE. WHERE ELSE IS THERE?

Jeremy had been born in Elsewhere, Louisiana. He had always found that somewhat ironic, yet it always made for a cordial icebreaker when he met someone new. It exuded a sense that he was ethereal, yet in the heaviest sense of the word. Once his audience finally realized that he was indeed from a place called Elsewhere, he had their attention. The tale circulated like any good story can. Often he was referred to just as Elsewhere. That was how his name got around, and it all had to do with the place that he'd run from.

The lore of his hometown was a good story and he would share it often.

"Well, these European settlers came to the area in the late nineteenth century. But the plot of land that they settled was later deemed undesirable. So the settlers sat around drinking, moping—moaning much like we do about real estate prices in Manhattan. The leader's wife, Dorothy, started nagging him to find a new place, something spacious but with less maintenance.

"And before you knew it, an expedition ensued. He wasn't

really looking for new land; he just wanted to get out of the house. But as fate would have it, they did stumble upon this fine piece of land that had been developed quiet nicely, thank you, by the Natchitochas Indians. Sound familiar?"

Jeremy would more than likely have hooked his listeners by this point. He was a good storyteller, adjusting the tale according to the crowd in attendance. This story was his calling card; his presentation flowed with ease.

"The lawyers were called in. Of course, back then, you didn't really have to pass the bar. No, it basically had to do with who killed the deer with the most points," he'd say, and after a pause, he always added, *"out of season."*

"Now, 'pending on the teller of the tale and who you're prone to believe, an agreement was reached to exchange the lands, and the Indians were to fork over some cotton and some rice, or ARROZ, for my Spanish-speaking friends. Well, the Natchitochas quickly agreed, for the settlers' land was of greater dimension than that of their own. When the blood had been put on the paper, they then pulled out the peace pipe—not because of the agreement; no, they just looked for any old excuse to fire up. And in retrospect, who could truly begrudge them that?

"Now this is where the buffalo chips started dropping. You see, the settlers felt they had received the better end of the deal, as is always sought when 'negotiating.' " Jeremy always put his fingers up to indicate quotes around negotiating, and all would laugh. He expected this, which always gave him an opportunity to sip his drink before he continued, seemingly enjoying the story as if it was the first time he had ever relayed it.

"The leader went back to his wife and told her that he'd found the perfect land, and a feast was had. Spam hadn't been invented yet, so I don't rightly know what they feasted on, but there was a toast, and it ended with his saying, 'Send dem Injuns elsewhere.'

"*But history is a bitch. No offense,*" *he'd say, looking right in the eyes of the hostess. He would wait for her to cup the wineglass under her chin with both hands and say, 'None taken.' And she always would. In these circumstances, he was always charming, his eye contact acute, yet not intimidating. He was no longer hiding behind the camera or his photographs. His legs crossed, drink in hand, he could work a room from a corner while becoming its center. Strangers would be less so, as is usual with an anecdote properly told.*

"*Time would soon show that the Natchitochas' newly acquired land was far more suitable for living than the settlers had previously believed. No, it wasn't as developed as now, and the bayou replaced the river that the Natchitochas had become accustomed to, but in the northern area of the boot, what it provided was a certain 'low land,' which tornadoes seemed to spare. In other words, a perfect place for a trailer park.*"

With that, he wouldn't give them an opportunity to laugh — trailer-park jokes were too easy — so he'd rise above their sounds, rolling toward the home stretch.

"*Unfortunately, again, 'pending solely on the teller of the tale, that landscape description could not be applied to the settlers' new land, something of which they were quickly informed a few moons later when a funnel and its curdled breath touched down without the slightest hint of consideration for the settlers' and the Natchitochas' agreement, literally blowing the settlers elsewhere. The name stuck, and like the most surreal real dream, the word* Elsewhere, *my hometown, can be found on the uppermost northeastern corner of the Louisiana map, a hop from Arkansas and a skip from Mississippi.*"

When Jeremy ended the story, he was certain that each and every one present would, at some point, pull out an almanac to point out Elsewhere, thereby passing on the story and his name. It all boiled down to name.

But Jeremy was far away from New York, the place where ugly ducklings from all over flew in hopes of becoming swans. Now he was back in that spot on the map that carried his story line.

Carl came out of the cockpit with a smile of accomplishment.

"Gotcha here in one piece."

"Seems as if you did."

As Jeremy's foot hit the ground, he couldn't help but look at his shoes. He remembered Mama B telling him when he was small that the road to the future started with a good pair of shoes and a nice timepiece. The shoes would get you there and the timepiece would make certain that you got there on time. But as he waited beside the plane for his bag, he knew that he was once again where time stood still for him, and all his shoes reminded him of was clutching, accelerating, and braking.

"So, you hit any holes in one lately?" asked Jeremy as Carl handed him his bag. Carl's face gleamed with fond recollection.

"Just sign here, Mr. Bishop," said the girl at the rent-a-car stand, popping her gum with a roll of the tongue and tuck of the cheek.

"My father's Mr. Bishop," he said, attempting a joke.

"Yeah. Whatever," *Pop, pop.* "Visiting family?"

Pop.

"Yes."

"I figured as much. We don't get many people renting cars here."

"Well, sorry to disturb you."

"Um-hum." *Pop.* "Sign here. Press hard and don't take the pen. You can have the blue BMW. It's just right out front."

"Do you have anything less fancy?" he asked.

She took out the chewing gum and rolled it between her fingers, then threw it over his shoulder out the door. "Got that silver Olds, but it's the same price as—"

"That'll do nicely."

With the key and his luggage in hand, he walked toward the

Oldsmobile. The past clung to him like an old sweater that only a moth could love. All those stories flashed before him. It wasn't that he'd forgotten them—he'd just put them away like undeveloped film in the bottom of a drawer.

He sat in the Oldsmobile. He adjusted the rearview. He took off his sunglasses and pinched the ridge of his nose to distract the pressure that huddled behind his eyes. His stomach churned as the engine turned over to a start.

I can see the picture in my mind.

The house I was raised in was home, not only to me but to many over the years. It was a wonderful place, with rose gardens and pecan and peach trees adding flavor to the surroundings. Other vines shared the earth with the roses, and I watched with anticipation as the fruit they bore turned from green to yellow to tomato. On more than one occasion, I picked them prematurely so I could line the kitchen window with them until the sun had had its way, finishing its shading.

The house was scraped with teeth of steel and painted white annually, long before the Southern summer's stifling breath could curl its coat.

The trimming was usually a color that complemented the redbrick supports that kept the house leveed from flood, flash or otherwise. This was the main house. The rental house, which was on the other side of the driveway, and the utility shed were painted the same colors.

The yard was always well maintained. I would sit watching Mr. Henry pull up in his red Chevy truck with all the necessary machinery that it took to turn a mere yard into a lawn. His son always tagged along, but we never spoke, and as far as I knew, "Mr. Henry's son" was his name. He would look at me as I peered at him through the window while his sweat beaded to a fall and worked its way into our soil. I would smile at him, looking for some sense of mutual acknowledgment— acknowledgment that would never come.

How I would have loved to go out there and join them, for it looked like the most fun in the world, or at least in the neighborhood. Father and son working together. But allergies kept me at the window, smiling while the lawn was cut and edged and raked until every blade was placed just so. When the driveway was swept, the two would pack up and be on their way, together, and from my vantage point, that said volumes.

I remember . . . not the day, not the week, not the date, which were all mere geometric squares hanging on the kitchen wall to me at the time. Julian or Gregorian meant nothing, for when I looked at the calendar, no names did I see — just space divided.

What I remember is this. I was five years old. He drove into the driveway in a blue MG. The top was down, and I could see him sitting in the car like a Jack-in-the-box. I'd seen the man before. I remember that. But he seemed no different from any other man I'd seen at that point in my tenure. I wasn't one for strangers; though to most, the word tends to mean any unknown being, to me it meant men.

Each time the man had previously come by, he'd looked at me as though I was something to behold, from afar, peculiarity encased. And I suppose with my eyes, older than my years, I looked at him in the same light. This man and Mama B and Aunt Jess would have a jovial conversation like a published chorus memorized. He'd look at me again, and when he finished talking to Mama B, he'd take one last assessing glance at me, then disappear.

That's what I remember most about this man; he didn't just leave, he disappeared.

I never thought of his absence; truth be told, I thought even less of his presence when he returned. He was just a man like many, if not any.

The sun was friendly and the birds played for as long as their wings pleased, before taking their rest along the charged wires between the oak poles with iron limbs for workmen to climb. I used to wonder how those blue and black and red birds could sit there without flinching, as I had already been forewarned of the danger of the currents. I wondered who taught them that it was safe, before that first time they perched there.

The MG looked like a toy to me, so to see a man in it made it seem queer. It was loaded with boxes. I thought perhaps the man had been coming by to talk to Mama B about being a contestant on The Price Is Right. But this man wasn't Bob Barker. No, this man's skin was dark because it was dark, not due to California sun or makeup of the same name.

"Patience?" yelled Mama B into the house from the porch.

"Ma'am?" I said, coming out to join her. I'd been stationed at the window behind the sheer curtain, an exhalation away from the glass, looking out as the old people did when the weather wasn't pleasant enough to man the porches, yet curiosity beckoned. I always came with a steadfast quickness whenever Mama B called, always eager to be near her side for the slightest of reasons. If I could have crawled inside her, I knew I would have.

"You're gonna go for a ride," she said, like it was the best thing that could have ever happened to me. I refused to believe it would be.

I saw the tiny car parked in our driveway and the man standing outside. His hand held open the screen door, but he didn't come in. I was certain that he would be reprimanded for allowing the flies to find new refuge, but the expected "Were you raised in a barn?" never materialized.

Three—that I managed to count—shooed their way by him, and I thought of the yellow butterfly flyswatter hanging on a nail on the porch wall. I thought of how I would take pleasure in using it to hunt each and every one of them—a fly safari. I thought of how I would cup them in my hand when they refused to realize that glass was deceiving and no matter how *they buzzed with frustration, they would have to find another way out. I thought about how difficult that would be without their wings, wings I'd place in a little box where I collected various gewgaws for momentary amusement.*

I thought of everything that I could, everything except this man.

He raised his pant leg a hem and placed his foot down on the third step, the last before entering or the first to leaving, 'pending on where the moment found you standing. I marveled at him, for his legs were so long that he could cover all three with one forward swoop, while I had to take one step at a time. Unless, of course, I was playing superhero and jumped down from porch to ground. But I wasn't playing at that instant and I had no intention of jumping toward him.

The man rested his elbow on the leg that was rested on the step and gave me that same look he had given me during every other appearance he made. But this time, the look seemed to say, This is your lucky day. *I was then convinced that he had been sent by Bob Barker and a "Come on down" was awaiting us. How else could the toy car and the boxes be explained?*

Aunt Jess came out of the front of the house and joined us on the porch. "That's a snazzy little car you got there, little brother. I see money still burns a hole in your pockets."

"Man's gotta get around," said the man.

"Now, I know after all I've done for you, you could at least take me for a spin around the block before you head off."

"Jess, you know you just too big and fine to fit in this car. I'd have to unload everything just to squeeze you in it. Besides, you wouldn't want to be seen with the likes of me. I might scare off all the other men."

When the man said this, a sugar-cane smile wrapped around his lips, leaving Aunt Jess with her own grin to chew on.

"Today, I'm takin' my boy for a spin," he said. "Then I gotta hit it. I don't know when I'll be gettin' back home, but

I promise, the next time around, the first ride belongs to you, and it'll be the ride of your life."

"Umph. That's what every man in town likes to say, and you ain't no different than any of them. I gotta change their diapers just like I had to change yours, and the only thing you get out of changing diapers is piss in the face."

They both laughed a holiday laugh, and Mama B chimed in as she looked at the two of them as though disagreeing with their repartee, but her jiggling shoulders showed otherwise. I was then convinced that this was one of Aunt Jess's "men friends." I'd seen them often in the front of the house, where her room was off the porch. The men systematically looked at me in the same way that this man did, question marks replacing their pupils. Sometimes I'd hear them say, "That your boy?" But none of them had ever called me "my boy."

"So, you ready to go for a spin?" the man said, returning his eyes to me. I backed behind Mama B, one hand to my mouth, the other clenched, knuckle white, to her apron. "See there? You're mama-ing the boy too much. That's no way for him to act."

"You haven't had any complaints up to this point, and now you wanna judge how we're raisin' him?" Mama B spat out the words so quickly and quietly that I could hardly make them out. It sounded like salt—not sweet—butter would sound if you could hear it melting. "Now, Patience, you're gonna go for a ride."

I didn't move from her side. I just looked at the man, but never at his face. I kept wondering what he had in mind and why today of all days I was having to go with him—this stranger. I wanted him to close the screen door. Yes, more than anything in the realm of my five-year-old world, I wanted him to get back in his toy car. I wanted him to go.

"I don't blame him." Aunt Jess laughed. "You ain't a sight for sore eyes. An eyesore, if anything." She turned her words to me. "J? Go on. I know he looks strange, but he won't hurt

you. You just gonna go for a nice ride in his car. He's just gonna talk to you is all, spend some time."

I looked at Mama B and her face delivered a sign that was meant to make me feel at ease. In her, I saw the eyes of countenance, saying, *Gwon, now. It's awright.*

Though skeptical, I released the handful of fabric, but the wrinkle of my grip remained, a sign of my steadfast resistance. I walked toward the place where the screen door would have been pressed shut had it not been propped open by this man. He put his leg back on the ground, and as I attempted to take my first step, he grabbed me, removing both of my feet from the porch's floor into the air faster than an old soul could digest or appreciate. The abruptness of it shocked me, for I felt no stranger should be so familiar.

I began to cry as he took me toward his car, certain that he was taking me away, never to return to Mama B and Aunt Jess. I culled the catacombs of my then small mind in search of my misdeed. He had me hooked between his arm and hip, while my legs ran on air. We never made it past the front of the car. The dent in the MG, a product of my struggle, captured his attention, and he put me down to better inspect the hood. Unlike a fish tossed back, I didn't scurry, swaying to the depths of safety to let my pulse settle and reflect on my newly gained freedom. Instead, I sat there in that very spot on the driveway where he'd put me down, ignoring the heat from the concrete that penetrated the fabric between me and it. I cried for a while more to assure him that I meant it and this was no childish ploy to get my way.

Things were said.

He said some things.

Aunt Jess said some things.

Mama B said some things. And though those things said had to top my high wailing, it was as though water filled my ears as well as my eyes, making all words completely muddled, simulating the voices of the adults that Charlie Brown always *spoke to. Still, I hoped the results of those things said would be that I wouldn't have to go with this big man in the toy car.*

When he popped the dent out of the hood, he looked down at me sitting there, beyond budging. He shook his head and brushed his hands together as though washing himself of me. I knew this gesture to mean I just don't know or Uh, uh, uh. Either meaning suited me just fine, for both were my thoughts exactly.

He hugged Mama B and Aunt Jess. Again he turned to me, looking down while opening his arms in the way that grown-ups would when they hadn't seen a child in a while and wanted that child to run into their arms with joyous force. I was certain it was just another attempt to get me into the toy car. I didn't rise. His arms lost hope and fell to his flanks. He walked around to the driver's side and slid into the car. The car started. He put on his sunglasses, and I was pleased not to have to see his eyes.

My flash flood of tears stopped immediately, but I could feel their snail trails drying on my face. As I sat on the driveway, I knew this man was about to run me over, and because I had rather that happen than go with him, I refused to let another drop fall. I focused on the remnants of insects that coated the eyes of the car and I waited to join them, need be.

"Well, maybe next go-around," the man said, resigned. "I tried."

"I know, son. But you can't rightly blame him," said Mama B. "It's your own doing. As far as he's concerned, you're a stranger, and he doesn't take well to strangers."

"I know. I'm sorry. Thanks for taking care of him. I do appreciate it. I'll make it up to you one day."

With that and a smile as large as the toy car, he backed out of the driveway. When the car was in the street, he blew the horn, and to me it sounded like a single wail, similar to the high-pitched squealing I had done just moments earlier. I knew the toy car didn't want to go with this man either.

"I'll make it up to you," he shouted from the street. "I promise."

Mama B picked me up from the concrete and held me in her arms. With my head resting in the nape of paradise, I heard her whisper, "It's not me you gotta make it up to, son."

Aunt Jess joined us, jumping up and down making all kinds of ruckus. "Wave 'bye to your daddy, J! Wave 'bye!"

She started running down the driveway and along the yard. She continued to say sundry things, but the only word I heard her say was daddy. It was in that instant I knew that that man was my father, yet I didn't have any idea as to exactly what that meant.

When the car and the man had disappeared and Aunt Jess and Mama B stopped looking in the direction it had gone as if they expected it to reappear, my hand did go up and I did wave at the man who was my father. For me, it was hello and good-bye, yet he never saw the gesture and he would never know which it signaled.

All that remained of him was a puddle of oil that glistened to reflection on the white concrete near the very place he had placed me.

I wanted him to go and he had. But the stain remained there for years.

chapter 3

๛ ๛

When Jeremy drove into the driveway, Jess was in the yard talking to a young woman. Two small boys were running around at play and a third child was being held on the hip of the stranger. Jess stared at the unrecognizable car, but when she saw Jeremy's face behind the windshield, her face lit with life. When he shifted the car into park, her conversation with the stranger came to a halt, and the woman and the three children graciously took their leave.

"J!" screamed Jess, jumping up and down just as she had done when his father drove away. But, now, age and weight had encased her, and her movements resembled winter, in that they lacked spring.

He got out of the car and gave her a bear hug. The embrace was long and needed no words, for the feelings were more than sufficient.

"Why didn't you tell me what time you were coming? I would've come pick you up," she said, wiping the tears from her eyes.

"That's all right. I just rented a car. I didn't want to be any trouble."

"Just shut ya mouth. When have you ever been any trouble? Let me look at you. Turn around."

Jeremy obliged her. He held his arms out at his sides as he

circled for her inspection, in the way done only for close relatives—methodically, enjoying it as much as they did.

"Look just the same. Just the same," she said. "Still no behind,
I see. But handsome as ever. I don't know why you haven't tied
the knot yet. I guess the Northern girls don't have a lick a sense."

"More than you know, Aunt Jess. Besides, you never got married. If I remember correctly, and I quote, 'The only thing a man
is good for is mowing the lawn, and he gotta be supervised when
doing that.' Unquote."

They hugged again, robust laughter between them, and then
Jess said, "Well, I'm not one to be tied down. You know how
it is."

"Yeah, I sure do," said Jeremy, his focus turning across the
yard toward the woman sitting on the steps of the rental house.
"Who was that you were talking to?"

"That's the new tenant."

"But I thought you said that you weren't going to take on any
new tenants; that it wasn't worth it anymore."

"Well, it's just sittin' over there, and as the saying goes, 'Waste
not, want not.'" said Jess, pulling out her handkerchief from the
crevice of her breasts. She wiped her forehead, then fanned herself with the damp swatch of linen. If spring was at a standstill
in New York, it had all but skipped Elsewhere. The heat was as
palpable as the two bodies that stood in it. "But she is truly trying,
Lord bless her soul. It's just out of the goodness of my heart that
I let her stay on. She was just giving me a sob story about not
being able to pay the rent this month. I said, 'This month? Chile,
you already two months behind.' Just a mess. Fast as she wanna
be. Ain't but eighteen and got three children, all from different
daddies, and not husband one.

"I tell you, we haven't had a good tenant in that house since
Charles, and nobody will ever keep up the place the way he did.
It was a sad day when we lost him. A sad day indeed. You remem-
ber Charles, don't you?"

"Of course I do," said Jeremy, looking over at the little house as though he expected Charles to come running out to greet him.

"Well, you always been good at remembering things. Seem like we always losin' people around here. I guess I'll be the next to go."

"Now, Aunt Jess, you know you're gonna outlive us all."

"I sure hope you right," she said, bending with laughter. "Come on. Get your bags, and let's get out of this heat before we both pass out from the stroke."

Jeremy pulled his bag out of the backseat of the Oldsmobile and they walked toward the porch. The screen door was now glass and the porch had been enclosed, obstructing what used to be a view. The washer was now on the porch rather than in the kitchen. His head filled with every used-to-be as the unfamiliarity of a familiar place presented itself. Though he had moved on, it was hard to believe that home had.

Jess was more like a sister than an aunt to Jeremy. Aunts were supposed to be a sprinkle of salt shy of a mother, but he never felt that way with her. She let him nestle under her wing without complaint. She recognized his old soul, for she was the one who had found him there in his mother's arms, and that formed a bond that would always exist.

Jess was the only one of Mama B's children who didn't go to college. When Mama Bishop was helping her husband at the Bishop's Funeral Home, Jess became, in essence, a mother to her brother and sister. By the time they went their ways, she could finally begin to enjoy her own life.

"It's the oldest child's responsibility to help out," she had once told Jeremy. But she didn't see him as another child to raise. To her, he was just J.

She was his first love. He once asked her to marry him. She simply said, "If I married you, things would change and people would talk, so I think we better just let well enough alone." That was her motto: Let well enough alone.

Of course, Jeremy loved his Mama B, in the way that boys should love their mothers, but Jess was different. She had never had a child but had raised three. Now, living in the house alone for the first time, she was able to live as an adult while an adult, rather than being a child with adult responsibilities.

In her prime, Jess was known for "doing hair." Not as an occupation—she didn't charge—it was mostly just for "sistafriends" who came over to tell the latest tale of show or woe. On Saturday afternoons, the kitchen was transformed into a pseudo–beauty shop and the local women would gather around the stove as though at a quilting bee. Jeremy would often sit on the washing machine, dangling his legs, hanging on every word. And it was there where he first began to learn about men and women.

"Come on in here and let me get in that kitchen," said Aunt Jess to Gloria, one of her oldest sistafriends. It took me a while to realize that the word kitchen meant something other than where food was prepared. I soon began to realize that it was the area where the naps of hair lined the back of the neck. "Girl, I hope I don't get shot when I put this hot comb back here. You got more buckshots than a white man during hunting season."

"Jus' watch it with that comb, 'cuz you burnt me up som'n special last time," said Gloria. "I almost had a mind to ast for my money back."

"Well, if you would stay still longer than two seconds, I wouldn't burn ya. You really oughta learn how to talk with ya mouth and not your neck."

"Lemmetellyasom'n," said Gloria, as she moved her neck from side to side, like an asp ready to strike. These visits would produce the same familiar responses every week: the tongue-in-cheek quips, the rolling of the eyes, the slaps on the back, and the bass-drum stomps on the floor. And every week, I would patiently wait for those appetizers to be done, eager for them to serve up the meat and potatoes.

Gloria was pleasingly plump and wore her polyester bright and tight. When she walked, the sound of her thighs rubbing together served as warning of her approach. Anyone who ever came into contact with her soon knew that her favorite word was lemmetellyasom'n, yet she never would say exactly what that "som'n" was. Still, more often than not, she never had to say any more than that, for that usually said it all.

On that particular afternoon, she was getting her hair done for a date to a hop. A hop was another name for a dance, but for Gloria it meant "get my cover paid, then hop from one man to the next."

"Who you seein' tonight?" asked Aunt Jess, who knew it took only one question to get Gloria going, thereby allowing Jess to focus on the hair at hand. She had to be particularly careful when Gloria was in the chair, for her movements were never telegraphed.

"Lester Murphy. He split up with Cassandra Meriwether. I guess she couldn't weather the storm. Now he free to blow on over here to a real woman. Ow, Girrrl! Watch that comb now. Lester gonna be all back there, and burns ain't the least bit attractive. Unlessn, o' course, you that adorable little Janet Jackson on Good Times. She can make anythang look cute. But I was so happy when Willona adopted that child, 'cuz her mama was burnin' her up som'n bad. Just a mess. Told her to take her shirt off. I thought she was gonna press it or som'n. Hell, nah. She wanted to press her little ass. I was screamin' at the TV, tellin' her to run, but she act like she don't wanna hear nobody.

"And that Thelma. I love that show but can't stand her. She ain't ever leavin' them projects. She find a rich man, next thang you know, he broke and livin' in the project with her."

"She don't sound that much different from you," said Aunt Jess, cutting Gloria's momentum. "Last time I checked, you were livin' over in the projects."

"Lemmetellyasom'n."

"And if my recollection serves me, weren't you seein' some rich man you was callin' Mr. X?"

"Oh, him? Hmmph. He's history. His wife found out about us and I had to cut that off with a quickness."

"Now how did she find out?"

"I told her."

"Gloria, why in the world did you go and do somethin' like that? You just don't know how to let well enough alone."

"Listen, he was tryin' to have his cake and eat it too, so I had to blow out the candles."

"I know you didn't think he was gonna leave his wife. Besides, you shouldn't be messin' with married men anyhow. That's just asking for heartache."

"I don't mess wit' 'em. They mess wit' me," said Gloria, raising her arms and making the makeshift barber's apron that draped her seem to inflate like a ghost. "And a married man is the best sorta man, 'cuz you gotta mess wit' the ones that got as much to lose as you got. That way, they mind they p's and q's. But see, this one here was gettin' outta hand. Thought he could just call me up at any ol' time and get him some. Called up last Thursday, tellin' me to meet him at the palace. Wouldn't even come get me. I had to take the bus.

"The palace was his name for our meetin' place. 'A palace for a princess,' he said. Palace, my ass. A flea-'fested motel is what it was. Palace? Umph. I told him, lemmetellyasom'n. And he was like, 'Ah, now, baby, don't be that way.' I said 'Did you carry me for nine months? Hell, nah. So don't be callin' me baby. I ain't ya baby. If anythang, you the baby.'

"See, he had been tryin' to get me to spank him and tell him what a bad boy he was. Always danglin' some handcuffs around. I didn't need no handcuffs to feel like whoppin' him upside his ass. But that's how those white men is."

"Girl, you didn't tell me Mr. X was white."

"Well, he shole wahn't no Muslim. I called him Mr. X 'cuz he'd say, 'X marks the spot,' and I'd smack that ass again.

But do you wanna do my hair and lemme finish the story, or do I have to take my business and gossip elsewhere? You know Jackie been tryin' to get me in her chair at BeautyLand. She been dyin' to get in this hair. You always interruptin' somebody. Make me lose my train o' thought."

"Handcuffs," I said, leaning against the doorjamb between the kitchen and dining room. I wanted to hear the rest of the story as badly as she wanted to tell it.

"That's right. Thank you, baby . . ." Hearing Gloria say *baby* made me imagine her as my mother, and for an instant, I imagined myself climbing inside her, yet only for an instant. ". . . so I get to the motel and he's all stretched out, naked as a jaybird. I started gettin' undressed right away, 'cuz let's face it, it's hard to be 'fectionate in a cheap motel room unless you truly in love. Then you can call it charmin'. At least that's what white folks seem to think, 'cuz that's what he said the first time he took me there. 'Don't you think it's charmin'?"

" 'Lemmetellyasom'n,' I said, 'cuz I knew his wife was sittin' over there up in that big house they got on Lakeshore Drive. I know this 'cuz he drove me by there one time, and when one of his neighbors passed by, he made me duck down in the seat. He tells me, 'Sorry, baby. But we don't want nobody messin' up our good thang, do we?' Like that made up for shovin' my head down like I'm some kinda cheap hooker. Don't you dare say it, Jess."

"I wasn't gonna say a word," said Aunt Jess, sticking her tongue out at me. "I'm not even studyin' you. I'm just doin' your hair."

"Annnyway, he stretched out, naked and grinnin', and when my clothes was off, I walked over to the bed. And lo and behold, there was them handcuffs again. He had 'em hidden under the pillow. I grabbed 'em from him and I did say, 'You been a very bad boy.' He was like, 'Now ya talkin'.' I told him to shut the fuck up. Oh, I'm sorry, baby," said Gloria, turning her words toward me without missing a beat, as *though her apology were a part of the story.* "Miss Gloria's mouth is just too nasty to be 'round the likes of you. But you see, he thought I was role playing. But I meant it. I wanted him to shut the fuck up. Ow! Girl, that burnt."

"Then you need to watch your mouth. J don't need to be hearin' that kind of language," said Aunt Jess, waving the hot comb in front of Gloria's face, yet smiling at me all the while. Evidently, hearing about an adulterous handcuffed white man was one thing, but she had to draw the line at excessive use of profanity.

"I'm sorry, precious pumpkin pie," said Gloria, apologizing again. "I do get a touch out of line sometimes, but I'm just tellin' it like it happened. When you think I'm gonna say som'n nasty, just cover ya ears, hear?"

I nodded my head in agreement and, being the little character that I was, I placed my hands over my ears, and she smiled upon witnessing it. That kitchen was filled with smiles, and mine was the largest.

"I put one hand at one corner of the bed and the other at the other corner. Chile, he got so excited, I thought it was gonna blow up right then and there. I say, 'You like it?' And you know, he did. He'd been wantin' me to handcuff him for the longest. I walked 'round the bed a few times, teasin' him, then I got dressed and walked out."

But Gloria hadn't simply left him there. She'd gone down-stairs to the motel lobby, gotten on the pay phone, and dialed his home number.

"I said, 'H'lo. May I please speak to Mrs. X?' 'Cept I used his right name, 'cuz she woulda said that I had the wrong number. But I had her number, awright. She said, 'Speaking,' real sweet and homemaker-like."

Gloria went on to inform Mrs. X that she was at the Palms Motel in room 307 with Mr. X and that she thought that she should come over and join the party. After Gloria hung up, she went back up to the room, where poor Mr. X was wiggling

and screaming. She told him what she had done, and he was
less than pleased.

"I said, 'Why don't you be like the dog you is and gnaw
ya arm off?' Then he had the nerve to call me a cunt. Uh,
sorry, baby," she said, apologizing again for my benefit, but
she couldn't look at me this time, for the hair on the opposite
side of her head was being rolled. "I'm gonna wash my mouth
out the minute ya auntie gets through. But . . . he did call
me that . . . that word.

"White boys love that word. When they gettin' they way,
they act like a l'il leprechaun at the end of the rainbow. But
the second they ain't gettin' it, they call you a . . . that word.
I hate that word. But I laughed in his face. I thought that
he should realize that when somebody got you naked and
handcuffed to the bed and ya wife is on the way, you might
wanna try and say som'n nice. But that's just me.

"I sat down and pulled out a cigarette and he asked me
what I was doin'. I told him I was gonna have a cigarette—
what did it look like I was doin'? Speakin of, baby, go in my
pocketbook and get me a cigarette."

I went over to get the cigarettes with the speed of Shazam.
I didn't want to miss even a breath of the story. She asked
me to light it for her. Aunt Jess told her not to get me started
with bad habits, to which Gloria said, "Lemmetellyasom'n."

I lit her cigarette on the stove and passed it to her. This
wasn't my first time puffing. I'd taken the butts of Mama B's
cigarettes that were in the ashtray in the bathroom and smoked
them down until the brand name disappeared.

Gloria placed the Kool Filter King 100 in her mouth. I
watched it bob as she continued her story with just a slight
change in her voice, the words sliding out of the left side of
her mouth, Popeye style.

"I sat there wonderin' what in the world was takin' Mrs. X
so long. Chile, had it been a black woman, she would have

been at that motel so fast you woulda thought she lived down-
stairs—chil'ren in tow if she had to. Now that's how we are.
But this bitch was takin' her own sweet time. She probably
felt she had to dress for the occasion. I could just see her
pacin' through the house, screamin' at the maid to bring her
som'n on the rocks. See, white women cry first, then they pull
it together. Oh, no. Not us, honey. We whip that ass first,
then cry about it later.

"Well, finally, a little knock came at the door. I sat there
for a while, just 'cuz I could, then I got up like I was one of
those debbitants that you see in the papers. I walked over real
slowlike, then I put my hand on the doorknob and looked over
at him and gave a little Mahalia Jackson smile. You know,
that kind of 'lemmetellyasom'n' smile. I opened the door, and
there was a white woman, but it wasn't Mrs. X. She had sent
over her sister. The woman walked in, looked at Mr. X and
just turned up her nose like she smelt sour milk or som'n.

"She asked me for the key. Started gettin' all uppity about
it. She said, 'Listen, this has gone far enough. Just give me
the key and we'll forget this entire incident.' Ain't that som'n.
I got her brother-in-law handcuffed to the bed, and she gonna
try and act like she in charge. I said—"

"Lemmetellyasom'n," Aunt Jess and I screamed at the same
time, and all three of us burst into laughter. I laughed so
hard I thought I was going to be like Pookie Baker from down
the street and pee on myself.

"You know I did," said Gloria, topping our howls. "So when
the pigs showed up . . ."

"Shut ya mouth. Tell me the cops didn't come," said Aunt
Jess, putting the hot comb back on the flame.

"Yes, they did . . ."

When the story finally came to its conclusion, her hair was
done—just the way it always worked. She'd look in the mirror
on the window ledge over the kitchen sink and would always

be pleased with what it revealed. I always hated to see her go, but I knew I would hear more of The Adventures of Gloria next time, and the story would be bigger than the ones that preceded it.

Aunt Jess once told me that if losers were pearls, Gloria would put every local jeweler out of business.

Yes, Aunt Jess was like a sister to me, and so she let me share these experiences, but a disclaimer always followed. She'd sit me down, and as if it were a conversation like any other, she'd say, "It takes all kinds to make the world go around. Remember that. Try and see and hear as much as you can. But you—and only you—can judge what is right and wrong for you. But don't you ever judge people based on how you act."

I too had my hair done by Aunt Jess. On quiet nights when rain seemed to spit down rather than fall, I felt so comfortable sitting on the floor, my back pressed snuggly between her legs. I felt like I belonged there and could stay there forever. Nothing felt better to me.

After washing my hair in the kitchen sink, I would plug in the blow-up comb, placing the switch on high, picking it out until my hair stood at attention. Aunt Jess would then take the hair pick with the Black Power fist on the handle and begin pulling and patting. I could feel the coldness of the metal scanning my scalp before freeing itself to the sound of popping static.

Once my hair had been picked, to and fro, she would take a comb and begin working her way from the left to the right ear, parting the hair until the scalp presented itself. With the line from forehead to neck completed, providing an almost zebralike effect, she would three-finger the olive green Afro-Sheen pomade. Her fingers became driftwood wading down a stream of scalp. From there, the braiding would begin. How I loved the way the pull felt on my skin. She'd often ask if *the braids were too tight, but they never were; without head-ache, nothing is properly achieved.*

"Did you used to braid my father's hair?" I once asked her, still wanting to find a bond to bail the feelings I had toward him.

"No. He mostly kept his hair cropped. That was the style back then. Like everything else, styles come and go."

chapter 4

∼∼ ∼∼

"**A**re you hungry?" screamed Jess from the kitchen, the one in which she'd stood years before with him at her hem. "People been bringin' over all kinds of food, and I'm sure Carol is gettin' her fair share, too. Ribs, ribs, and more ribs. I tell you, people act like they don't know how to fry a chicken anymore."

Jeremy leaned against his old spot, the doorjamb between the kitchen and dining room. He stared at the microwave that looked so inappropriate for the old-fashioned, down-home kitchen he once knew so well.

"Or else it's ordered. People don't cook these days like they used to. Just go somewhere, buy it, then act like they done somethin' special. Times are just too fast. Too fast, I tell ya. Just can't let well enough alone. Good enough is never good enough. I say the day people in the South stop cookin' is the day the South dies. But I guess it's just too warm out to be standin' over a hot stove. No need in cookin' when somebody can do it for you and deliver it. In my day, we cooked and proud to do it, and you made them beg for the recipe."

He stood there looking at Jess and thought of his refrigerator in Manhattan.

Film. White wine. Bottled water. Vodka. Spoiled orange juice, from his last bout with the flu. Mayonnaise. Cherry Garcia Ben & Jerry's, as well as the rolls of pounds, francs, and pesetas for flavors on foreign shores.

As for food, he had nothing but white cartons of Chinese food, but he knew it was once Chinese food only because that was what he had ordered. Though he refused to throw the cartons out, he dared not open them, fearing what he might now find. At least the boxes gave the appearance of something there and seemed less pathetic than an empty refrigerator.

"That's fine, Aunt Jess," he said. "I ate on the plane."

"I hear that plane food'll kill you quicker than a crash."

"It's not so bad in first class."

"Well, my, my, my," said Aunt Jess. "You go on, boy. Just like ya daddy, always first class."

"Don't say that."

A hush broke the jovial mood like a sudden gray mist would, but Jess kept going as though she wanted to table the matter.

"One of these days, I'm gonna fly up there to New York City and pay a visit. Maybe go stand out there and watch the *Today* show."

"Oh, famous last words. You've been saying that for years, and I've been offering to buy you a ticket."

"You know I can pay my own way. It ain't about the ticket. It's the plane. They make me nervous. I'd be nervous on it, then I'd get to New York City and be nervous 'cause I'm in New York City. Then after that, I'd have to fly back and be nervous. No, sir. That's just a bit too much nervousness to call it a good time. I'll sit right here, thank you. But that don't mean you can't come down here more often. It shouldn't take a funeral to get you to come home."

The word *funeral* rang out like the town hall bell at noon. There was no denying it; this was no holiday. He was here to bury the dead—bury his father.

"So what happened?" he asked, grabbing a plate from the dish rack on the sink. He began to fill it with food. He needed some form of distraction, something to strengthen him or at least weigh him down.

"Well, he'd been sick for a while. His heart. Pressure just kept

41

shootin' up. But it was his kidneys that caused the real problem. Some kind of arthritis of the kidney. Chronic pyelonasomething or 'nother."

"When did they diagnose it? I mean, wasn't there some sort of treatment?"

"Don't nobody know for sure when he found out. You know how he was. Didn't tell nobody. Thought he could lick it on his own. He's a man, and men always been afraid to go to the doctor, and of course, he hated the hospital 'cause it reminded him . . ." Jess let her voice trail off. She looked up at Jeremy, but he kept filling his plate as if the open pots were fulfilling his need for answers.

She continued. "Then he had a minor heart attack. Ain't that the silliest thing you ever heard? 'Minor' heart attack. You ask me, ain't nothin' minor about no heart attack. A heart attack is a heart attack. I guess the only difference between minor and major is the sound that little machine next to the hospital bed makes. If it sounds flat, that's major."

Jeremy almost choked on his mouthful of baked beans. Though she hadn't meant to be funny, he couldn't control himself. He'd always loved her for her candor, and now he realized how much he'd missed it. It wasn't the cynicism calling itself truth that he dealt with daily in the city, it was observation in its simplest form. It was that truth that tightened his diaphragm.

"This ain't funny," said Jess, taking his plate from him and putting more pork ribs and candied yams on it. "Heart attacks ain't no laughin' matter. As sure as I'm standin' here, it'll kill you."

"I know, Aunt Jess."

"I told ya daddy, time and time again, to watch himself. But you can't tell men nothing they don't wanna hear," she said, passing the plate back to him. "Y'all just know it all. Now, you see, a little weight on a woman is fine. Men love a big woman, somethin' to hold on to durin' the lean months of winter. But see, women work. We constantly on the move; that's what we do. That's our job. That's why we can carry some weight.

"Men like to think they workin' just 'cause they makin' money. But what they spend most of the time doing is sittin', eatin', and runnin' they mouths, and don't none of those things cure cholesterol. Peck here, lunch, peck there, dinner party—next thing you know, they sittin' in front of the television watchin' the news with their pants unfastened, pattin' they belly like that's something sexy to look at. I can do without."

Jeremy laughed again as he watched Jess move around the kitchen with ease. Yes, he was home. And as much as things had changed, some things had definitely remained the same. During his meal, three helpings—"Don't be shy, now"—he found out that Carol was making all of the arrangements.

"She wanted to call you herself," said Jess, her voice finding decrescendo. "I told her I would break the news to you, but she was set about it, so I gave her the number to the studio. That's the only place I ever catch you, and you know I can't stand those answerin' machines. Seemed important to her, and I didn't feel like puttin' up a fuss. I mean, it's one thing to lose a brother—that hurts—but to lose a husband, well . . ."

Jeremy had never called Carol his stepmother. He referred to her simply as Carol. If the occasion ever came up that he had to mention it, she was always known as Carol, his father's wife. He rarely ever mentioned it, for he knew it would lead to other sentences, which ultimately led to his having to say that his mother died. In some cases, that would be enough and he'd segue to another subject; other times, the common "I'm sorry" would roll out of mouths feigning sympathy. He would always follow up with "Did you kill her?"

"She called this morning to see if I knew when you were comin'. I told her I didn't know exactly, but you would be here sometime today."

"How's she taking it?"

"She's hangin' in there. But you can tell it's hurtin' her. A lot of new questions she's got to field and answer by herself. Jessica is a free spirit; she seems okay, but she only lets you know what she wants you to know. But Jason is taking it hard. They were close."

Jeremy's question had been posed with sincere concern, but the answer spoiled his sincerity. He did his best to conceal the vile taste of envy. Others knew his father better than he did, and now, that was certain to never change.

"The family hour is Friday night and the funeral is on Saturday. You should really give Carol a call."

"I will."

"You want me to run you some bathwater?"

"No, I'm fine right now."

"Now, there's plenty of food if you want some more."

"I don't think I could eat another bite."

"Well, you sure could use it. That behind ain't gonna ever fill out if you don't eat," said Jess, patting his backside. "I'll leave it out just in case. There's all kinds of sandwich meat in the fridge."

He didn't want anything else to eat, but he continued to stand in the kitchen, looking for the slightest detail to take his mind away from his thoughts, away from the memories. Away was always familiar. His feet were planted firmly, but his mind was racing, and that proved more exhausting than anything else.

He walked through the dining room. It remained the same as it had when he was a child. The same plastic arrangement was centered on the table; the oranges and pears and bananas and grapes, forever ripe; the flowers, forever beeless. He walked to the Deepfreeze and opened it, allowing its cold breaths to scramble out, clinging to and tightening the warm pores on his face. Someone had been kind enough to fill it with the season's givings. He remembered the joy of quail and the stickiness of deer, now called venison. He remembered the purple hulls and the broadaxes, frozen still in plastic bags. He looked at the large butter container that now held homemade ice cream. Yes, everything was in its place, and in that, he felt full.

The pictures that lined the walls of the living room stared out at him as though he were seeing them for the first time. He looked closely, searching for some semblance of those moments in time. The Afro, the Caesar, the cornrows, the part, the fade. All his cuts drew out a different feature of his face, but the eyes were constant.

When he had brought home the proofs of school pictures to decide as to which package to purchase, Mama B would always ask why he never smiled. Though he'd say he didn't know, he truly believed that smiling would mean that he wanted his soul to be stolen. Yes, the pictures would eventually be purchased, but it would always be the smallest package. Rarely were images cut and exchanged with other students who had so meticulously wet the corners of the photographs to inscribe their names on the acetate in their best hand.

"You always could look good in a picture," said Jess, coming up behind him. "Don't surprise me one bit that you takin' pictures now. It always looked like you were lookin' at the camera, tryin' to figure it out."

"Yeah," he mumbled, "I guess that was it."

"I've always loved this one of you on the front porch," she said, pointing at the photograph with him wearing an Indian headdress. "You would skip and dance around like you had ants in ya pants when you were in that thing. I do believe you thought you could make it rain. It used to tickle Ma Dear to death."

"Yeah, I thought I was one of the Elsewhere Indians. You know, I tell everybody the story of Elsewhere. They eat it up."

"You always could tell a story. I remember those little poems you used to write, and then you'd make us come in and you'd put on a recital. Just as cute as you wanted to be."

Jess stood there beaming. She had always been his biggest fan, long before he ever thought he could do anything but long.

"And you know who's always askin' after you?"

"Who?"

"Ol' Mrs. Richards, over across the street."

"Get out," said Jeremy. "She's still alive?"

"Ol' as dirt and as hard too, but after Rascal died, she calmed down a bit."

Jess placed her hand in his, and he held on. Her skin felt like velvet, smooth yet worn by touch. He could begin to see her as older, but the thought didn't discourage him, for he was getting older, too. She started walking him to the back of the house. He was well over a head and a half above her now, but he felt like a child again; their steps, memory lane. How comfortable the side of his face had felt rested in the space between her hip and torso, as though that curve had been made for just that very purpose.

"You know he loved you, don't you?" she asked after a few steps of silence. "He was just like me, just like Ma Dear. Everything he did, he did for what he thought best. You know how it is on a young man tryin' to make it in this world. He felt he needed to make his mark, fill out his name. Well, he did it. Just like you're doin'."

"I know, Aunt Jess," he said, but the mark he remembered most was the one left by his father on the driveway.

The humidity of the day had ushered in eggplant-colored clouds, accompanied by a countering breeze that freshened the air. A storm was approaching. He sat on the porch, longing for the rain to pour. He had missed the hovering thunderstorms that made the house—the neighborhood—still.

No radio. No television. No lights. No nothing. Just silence, as though noise would anger the storm. In actuality, it was just a moment to appreciate the power of the unknown, a moment to hum, to rock, to sway. A moment to reflect or contemplate. A moment of one's own.

He scanned the street. It had changed—or maybe it was the same and it was his view that had gone askew. The lawns didn't seem as well kept and the houses lacked new coats of paint. It appeared barely a generation away from ghetto. Farther down the street, spray-painted images tagged the particleboard that concealed any window view.

The street had definitely changed.

Several attempts had been made to get Jess to move. She could afford it. "Just 'cause you can afford something doesn't mean you need it. Every old sock needs an old shoe," she had told Jeremy when he brought up the matter a few years earlier. She was set in her ways, comfortable in her surroundings, the life she knew. She had run this castle, and now she reigned over it. This was her family plot.

He sat on the porch swing waiting for the first lightning to break the sky. He closed his eyes. He took a deep breath. He smiled. It was a good, long smile, much needed.

No radio. No television. No lights. No telephone. No explanations. Just appreciated silence.

Memories of all the houses were coming back to him like a developing wash, and he viewed them how they once were. Even some of the ones that had looked pretty back then had also held ugliness that willingly presented itself once he left the safety of his own; protection extends only so far.

When he opened his eyes, he stared at the house directly across the street. It suited the dark clouds that seemed to frame it. He couldn't believe she, of all people, had asked about him.

In my mind, Mrs. Richards was a witch. Not a Bewitched sort of witch, but a witch of toil and trouble, boil and bubble. I was convinced of it. She had a mole on her nose just like every wicked witch that I'd seen in books. It was because of her that I preferred the back porch to the front. She had an eagle's-eye view of our front porch, and what it saw in me wasn't the least bit appealing. I was not known to her as Patience or J or even any of my many names. To her I was . . .

"Ugly boy?! I say, ugly boy?" she said, hollering across the street. "You hear me callin' you, don't you? Don't be ig-in' me. Come here and don't make me call over there again."

I looked around to see if Mama B was around. She wasn't. I knew I was allowed to go anywhere in the neighborhood without telling her of my whereabouts, but Mrs. Richards was the devil incarnate and I would always look for any excuse to stay far afield from the witch's den. Yet not wanting to "ig" an adult, I took off my Indian headdress, stepped off the front porch, and crossed the street.

"Look at you. Just spoiled and rotten to the core. Now, ya daddy was sweet as blackberry pie. But you, you should stay in the house when I'm out here. You're ruinin' my view."

I didn't say anything. What could I say? I remained focused on her mole. With every word out of her mouth, it seemed to grow, engulfing her face and dissolving her words into dust.

"But since you're here and have already ruined my day, why don't you make yourself useful and go run 'round to Scalia's and get me a pound of salt meat. And tell that cheap Scalia not to be stingy, either. And don't you be tempted to steal any of this here money. I know exactly, to the penny, what a pound of salt meat feels like and I'll be wanting my change."

I too wanted change—a change of venue. But my feet were planted on her porch as I watched her reach into her bra. I assumed she had a potion in that knotted handkerchief, but she took out some coins and handed them to me. Her finger-nails touched my palm, which I was certain would be covered with warts before day's end. The thought made me drop the money onto the porch.

"Lawd have mercy—clumsy, too. What are you good for? Don't just stand there like you're simple; pick it up. Every cent of it. Did any fall through? If it did, you just better get ready to crawl ya narrow l'il behind right under this house and get it, and I hope a snake bites you. It'll serve you right."

I kneeled to pick up the coins, luckily all present and ac-counted for. I could see the lifeless knee-highs draped atop her laced-up black shoes, wrapping her ankles like unwatered plants hugging the sides of a pot. Before I stood, I looked back across the street toward home, hoping to see Mama B, hoping for some rescue. Hope . . . lost. "Make ace, now. I don't have all day."

With that, she vanished behind her screen door. I tried to peek in, see the crystal ball and the broom that was for more than sweeping, but the door slammed too soon, yet not soon enough to keep Rascal from shooting out.

Rascal was the only black cat in the neighborhood. His fur shimmered blue in the light and his piercing blue eyes gleamed like marbles any boy would want to add to his bag of prized possessions. Rascal was as sinful as yard work on Sunday. Let him cross your path, and you were certain to die that very day, unless you found another path or walked backward on the very same path that led you there.

Rascal looked at me. Though a sound came out of his mouth, I couldn't rightly call it a meow. It was a cacophonous yelp. But what I heard was "What are you waiting on, ugly boy?"

No matter Mrs. Richard's request, I would have fulfilled it, for refusing an elder, evil or otherwise, wouldn't be acceptable, but Scalia's market was actually one of my favorite places. It was only a few blocks away, and I often went there to buy Mama B's cigarettes. It looked like a house on the outside, but once you walked in, there was no doubt that it was a store. The jars of pink pig ears and snouts always startled me, but those candy-filled jars that lined the counter were a sight to see. All the colors and flavors were well represented: the lemon drops and the jawbreakers and the Blow Pops and the gumdrops and the Tootsie Rolls and the Now & Laters, all in that order. I had memorized their placement, so even when the store was blocks away, I could still imagine it all. I remembered those jars like I remembered my name.

Mr. Scalia and his wife were always pleasant. Though they didn't live in the neighborhood, they were neighborly. They knew everyone by their first names and could recall which children belonged to whom. Credit was given without any signatures. They knew you weren't going anywhere, and neither were they. They were good people serving good people, and more than that, they always gave me a peppermint stick to stuff inside my dill pickle, "no extra charge."

"Jeremy! Hey, how are ya?" shouted Mr. Scalia from the back of the store. He was always in the back where the meat counter displayed its goods. His hellos were always yelled, for he was hard of hearing and had difficulty monitoring his volume. Mrs. Scalia tended the front of the store, where everything else was displayed.

"I can't complain," I said, taking on my adult tone. "But I must say we're due for a change of weather. Wouldn't you say so?"

"You can say that again," Mr. Scalia said with a chuckle.

"I said we're due a change of weather!"

Mr. Scalia's chuckle turned into a full-fledged laugh. He was always good-natured and ready for any exchange. I had learned from listening to adults that when in doubt about what to say, talk about the weather.

"What will it be today?" asked Mrs. Scalia. "A jawbreaker, maybe? Or is it a honey bun?" I had to swallow long and hard at just the sound of the words. But my mouth quickly dried when I remembered why I was there.

"Mrs. Richards sent me over for a pound of salt meat."

"Sal! Richards wants her salt meat," screamed Mrs. Scalia, adding, "Don't skimp on it, either. She knows, to the penny, what a pound of salt meat feels like." When she said that, she looked down at me and winked as though she had magically been standing with me on Mrs. Richards's porch.

When the meat was wrapped in the butcher's paper, which was taped shut, Mr. Scalia brought it up to the front of the store and placed it on the counter. I handed Mrs. Scalia the money.

"So what can I get for you?"

"Oh, no, ma'am. I'm just running an errand for Mrs. Richards. But thank you kindly." My eyes must have said otherwise as I drank in that counter and its young-heart pleasures.

"Well, let me see. What's this here?" said Mr. Scalia, bending down toward me.

"Where?" I asked, certain that a mole or more had grown on my nose and that he was amazed by the sight.

"In your ear there. Do you see it, Mama?"

"Why, I do think I see what you mean," said Mrs. Scalia. I began to get nervous. I had cleaned my ears—behind them, too. Of that I was certain.

"Well, how in the dickens did that get in there?"

When Mr. Scalia removed his hand from the side of my face, there appeared a box of Alexander Grapes. He shook the box, and within it I could hear the little purple candies jumping to attention. He slipped the box into my shirt pocket, I gave him my "Thank you very much"—still amazed—and I was on my way.

As I left the door, I wondered if that contraption in his left ear was some sort of candy that I wasn't familiar with, but I knew I would never ask or reach there to find out.

I kept the box of Alexander Grapes tucked next to my heart as I walked, holding the salt meat next to my hip. My step was lively, my mood the same. When I got to Mrs. Richards's door, before I could even raise my hand to knock, she appeared behind the screen, which was almost as black as if it had been charred. She looked at the package of salt meat and latched her door.

"Where's the bag?"

"Ma'am?" I asked, somewhat confused, knowing that meat always came wrapped in white butcher's paper, not in a bag.

"The bag! The bag!" she said, her voice rising an octave. "You take that right back. When you buy something, they should put it in a bag. You hear me? And you tell ol' Scalia

I know what he's up to and I'm not gonna have it. And what . . . what's that in your pocket? Candy? I know you didn't buy nothing with my money."

"No, Mrs. Richards. Mr. Scalia pulled this out of my ear. He said I could have it."

"No sucha thing. You go right back there and . . . and get me a bag, and you give that candy back. Ain't nothing free, you hear? Ain't nothing free."

I again turned around and headed off her porch. I again looked across the street, almost praying for any sort of interception, to no avail. I took the same trek that I had just minutes before, but this time the pleasure of going to Scalia's wasn't found in my hot-water stride.

"Why, Jeremy. Did you forget something?"

"No, Mrs. Scalia," I mumbled, putting the package of salt meat on the counter.

"Was something wrong with the cut?"

"Could you put it in a bag, please?"

I didn't look up at her. The store seemed to shrink around me. I focused on a spot on the floor and wished that I could fall into the crevices in the dirt that had accumulated there that day.

"Is there something wrong with the cut?" asked Mr. Scalia. Mrs. Scalia didn't scream back to him; she just shook her head in the negative and he went back to his business. She picked up the package of salt meat and placed it in a brown paper bag.

"There you go," she said with an involuntary smile as she handed the bag to me. But this smile was a different one than what I was accustomed to from her. A true smile never reveals wrinkles, yet I could see every indentation. "You have a nice day, Jeremy. We'll see you soon."

"Have a good one," screamed Mr. Scalia. He hadn't heard word one, so the smile on his face was true—so much so that it drew me toward him. As I walked through the store, the

products that lined the shelves became a blur. At last, I found myself in front of the various meats, and my stomach turned. I pulled the box of Alexander Grapes out of my shirt pocket and, on tiptoes, placed it on the top of the counter. Without a word, I walked away. I don't even remember walking to the door. I didn't look at Mrs. Scalia as I passed, but I knew she was looking at me.

All of this because I had failed to get a bag for a purchase in a world where nothing is free.

I needed air.

Mrs. Richards was on the porch, broom in hand. I handed her the bag without comment. She said, "And the change?" I dug into my Lee jeans and pulled out the change, handing it to her. It was a nickel and a penny, and no thanks replaced it.

"Patience? Come on home and stop bothering Mrs. Richards. It's time for lunch." Finally, a familiar voice interrupted us. It was Mama B calling from across the street. "How you doin', Sadie? I hope you stayin' cool."

"Just fine, praise the Lawd. L'il Jeremy here was just runnin' a quick errand for me. Just the most precious l'il thing, he is. I hope you don't mind," she said, rubbing her hand over my head, making my entire body cringe and shingling my skin. Rascal wrapped himself around my leg. I was certain this was hell.

"Not in the least. I'm sure he was pleased to do it."

"When is that son of yours coming home? I love that boy. You tell him I thought of him when next you speak."

"I sure will. Come on, Patience."

I walked across the street, unpleased and pleased with each countering step. I wanted to run, but the energy was zapped from me. I remained steady, concentrating like I did when walking on the rails of the railroad tracks and didn't want to fall into the crevice of nowhere.

"That was very nice of you to help out Mrs. Richards," said

Mama B as I climbed the steps back to the love and comfort of home. "I've got a surprise for you."

It was hardly a surprise. My favorite treat was cornbread. Whether with Foremost milk and sugar or just plain, I loved it. Mama B would always cook one tin for supper and another small one just for me. This wasn't the Jiffy brand that I would later discover and cherish. No, this was homemade, and when it came out of the oven dusk brown, I was rarely patient.

"Give it time to cool," she said, pushing me away from pouncing. "People always so anxious to cut into things they forget to let it cool and that's why it falls apart."

I loved that cornbread. With each bite, the outside world dissolved.

chapter 5

࿇

"**J**!" screamed Jess. "Phone."

Jeremy returned to the present from his beclouded state. He knew it had to be Carol. He wasn't ready to deal with her just yet. He walked in from the front porch and took the phone from Jess.

"Hello?"

"J.B.?"

"Yes," said Jeremy tentatively, not quite wanting to commit to a conversation.

"It's Paul. What up, playa?"

"Nothin' much, man. What's up with you?"

"Chillin'. Just chillin'," said Paul after a pause, like the static of changing channels. "Sorry about your ol' man."

"Thanks." Jeremy knew if anyone was truly sorry and could understand, it was Paul.

"So listen, some of the ol' posse is gonna hook it up tonight. Why don't you stop by? That's if you ain't too high and mighty to chill with common folk."

"Who is that?" said Jess, mouthing the words.

"It's Paul," said Jeremy, placing his hand over the receiver. When she realized her intrusion was safe, her voice returned to that of home.

"Then tell that boy can't he say hello to nobody when he call their house? And I know he knows better than to call when it's

55

about to storm. Y'all both gonna get 'lectrocuted, and I just can't take that right now. It would ruin my day."

"J.B.? Ya there?"

"Yeah. Sorry, Paul. It was Aunt Jess getting on your case for not speaking."

"You tell that sexy thang I said wassup, and as soon as she says the word, I'll divorce Beth and marry her."

"Paul says hey, and he's still waiting for your hand in marriage."

Jess blushed, but she hid behind it. "That's all fine and dandy," she said. "He can't even come over for a sit-down with nobody but can call up when it's about to storm. Get off the phone."

"Paul, I gotta get off before Aunt Jess body-slams me."

"What about tonight?"

"I don't know. With everything . . ."

"I know you ain't punkin' out on me."

"J," yelled Aunt Jess, "I'm not gonna tell you again."

"When and where?" asked Jeremy in a whisper.

"Around eight. I'm still in my folks' old place."

"I thought you were gonna move out of there."

"Get off me. I'll move when I'm ready."

"All right, but I gotta get off before Aunt Jess comes back in here."

"Take the phone on the porch."

"The cord won't reach."

"Don't tell me she doesn't have a cordless."

"She just got touch-tone a few years ago—get off her. Not everyone needs to move so fast."

"J!" screamed Jess on the last echo of a clap of thunder.

"I'm off," Jeremy yelled back, then said into the phone, "I got to jump in the tub. Is eight-thirty cool?"

"Bet."

"I'll see you then."

"Peace."

"Ciao." Jeremy hung up the phone and a smile came over him. As though an angel had seen it, the sun began to break through the clouds, making the drops of rain seem like crystals falling from the sky, presents from God to the devil's wife during her beating.

Paul had been Jeremy's best friend at Bonaparte High, perhaps his best friend ever. He was his "ace boon coon," as Paul would say. Where he picked up the phrase, Jeremy never knew. Paul married Beth right before Jeremy left for New York. Jeremy had served as his best man, for Paul's brother couldn't.

"You goin' over to Paul's tonight?" asked Jess, opening the window to catch the fresh breeze that blew the passing clouds.

"Yeah. He sends his love."

"Well, you're gonna miss Gloria. She's comin' over later."

"Tell me you're not still doing her hair."

"No, more like her wigs." Jess laughed. "But she's still up to her ol' ways. Got a wig for every day of the week. She calls them her mood wigs, like that ring you used to have."

"I do hope her mood doesn't warrant handcuffs."

"Well, she claims she gave all that up. But lemmetellyasom'n, she is still crazy as all get-out. Says she's lookin' for Mr. Right. I always tell her there's a difference between Mr. Right and Mr. Right Now and they ain't no way related."

Laughter filled the house as it had years ago, and a hug followed.

Yes, he was home.

"Do tell Gloria I said hello."

"She'll hate that she missed you. But she'll probably be at the wake—the funeral for sure."

There was that word again. It hung over the room like an echo bouncing off air, reaching the depths of his ears. No matter how he tried to avoid it, the word was always on deck, ready to be thrown at him, ready to coerce, prod, and remind.

"Since you're headin' that way, you really oughta stop in on Carol, and I know Jason and Jessica will want to see you. They have just shot right up. Kids just grow up so fast these days.

But they are always braggin' about their brother that lives in New York."

"I will, Aunt Jess," he said, then slightly lower, as though to himself, "tomorrow." But her ears were as acute as her stare.

"I don't know why you wanna put it off. You're gonna have to deal with it at some point."

"I said I'd go over tomorrow, okay?" His tone was the one he might use with his assistant, Doug—sharp and direct, one that often left him having to apologize moments later because he was indulging himself by hiding behind a facade of power. But that tone didn't wash here. As soon as the words left his mouth, both Jess and he started to grin. Jess looked at him, her eyes wide open and her head cocked back to the farthest angle her vertebrae would allow.

"I don't know who you think you're talkin' to, Mister," she said, placing her hands on her hip. "You must be talkin' to yourself. I know that, 'cause I'll fatten that lip for you if you don't watch out, and I won't think nothin' of it."

The two began to dance around like boxers sparring, but their dance ended with yet another round of hugs and laughter, as it always had.

Jeremy began to unpack. He hung up the two suits that he'd brought with him and pulled out the shoes that would be called dress shoes here. These were normally clothes he'd wear to social functions in New York, where cigarette smoke would cling to the fabric like an invited guest. Now they were to be worn for something else.

He went into the bathroom and began running the water in the tub. Though the room had been given a facelift, the hot water bottle still graced the back of the door. Baths were a major part of Jeremy's life. New York was a shower sort of town, everyone always on the move, keeping stillness at bay. He'd grown up taking baths, and even in New York, he continued to take them.

Running the water and waiting for the tub to fill was like therapy for him. It gave him permission to slow down.

He poured the Epsom salt into the water, then began to undress as he watched the crystals dissolve. Standing in front of the mirror, he looked at the stretch marks that formed semicircles around his ass, stretch marks that had been born in this very house, witnesses to his growth spurts.

He stared into the mirror as though seeing himself for the first time. No, he wasn't an ugly boy. Though he had often thought so, time had shown otherwise. Over the years, many had said that he should be in front of the camera, but he always let the subject pass. Yes, he had grown to be an attractive man; still, he never assumed others found him so. He'd often say, "Attraction is something others bestow upon you; you don't bestow it upon yourself."

Flirting was wasted on him. He had no pick-up lines to proffer. He needed controlled environments and proper introductions before he could feel comfortable enough to engage in conversation. It would be safe to say he was shy. Growing up, he had loved women so much that he was now intimidated by them, afraid that he could never live up to their expectations—or worse, that he would be what they expected and would hurt them.

He never understood when the word got back to him that others considered him unapproachable. He found it even more disturbing when he was passed over for the person who was always smiling. He never trusted people in New York who smiled all the time. To him, a fashioned smile meant exactly what "How are you?" had come to mean—nothing. Nothing other than something to hide behind, something to fill the moment.

Sexuality was an oblique word at best. He had explored, like many his age who refused to be defined. When he found himself with friends at "gay clubs," he felt at ease, refusing to allow himself to preface his presence by using the common excuse "I'm just here for the music." When pressed on the issue, he would often comment that he had a "working relationship" with his sexuality. But like anyone, Jeremy longed to be loved.

But all that didn't matter now. He was away from the city and her alluring ways that posed as freedom. Away from the parties and the doubts. Away from the insecurities that always find themselves cuddled between the rival siblings, acceptable and unacceptable.

With the eye of a diamond cutter, he scoped the top of the hamper against the wall that had always served more as a table than a repository for dirty clothes. The hamper was covered with glass jars of dusty, multicolored bath balls and shell-shaped soaps; an open box of Arm & Hammer; and the plastic container of body powder. He picked up the top of the latter to find the paper torn back and the powder puff sitting solo, where the scented dust used to reside. It had been there in that very spot for years, next to the mineral and cod liver oils and all the other accoutrements for pampering or treating ailments.

The nakedness of bathtime freed him to relax. He continued his perusal. He opened the bathroom closet where the towels were and saw the electric foot soaker, cotton balls, Q-Tips, the tin of prophylactic tooth powder that looked like an antique, and the Vicks VapoRub, which Mama B had warned could be administered only at night, for if it was used in the daytime, it induced a fever. He winced when he saw the bottle of castor oil—every grandmother's worst threat when children attempted to lie about having a stomachache in hopes of staying home, away from the exam for which they were ill prepared.

When his bath was full, he sat on the shoulder of the peach-colored tub for a moment, running his hand through the water. He had done this many days in this very room, when he hadn't taken a bath at all. He was always a clean child and went through a phase when taking a bath seemed unnecessary. But in Mama B's house, "cleanliness is next to godliness," and missing a night could have easily violated the eleventh commandment as far as she was concerned.

Still, he would sit there running his hand through the water, letting his imagination run while making splashing noises to make it sound as if he were in the tub. When a suitable amount of time had passed, he'd pull the stopper, splashing water on the bath towel to make it an accomplice to his deceit.

All these memories came back to him in cinematic detail. He stood up and placed his right foot in the tub. His foot quickly adapted to the temperature, and the left one soon followed. He stood there jostling the water with his feet before he slowly slid his entire body down, allowing his breath to escape through every pore of his skin.

Though at one time he could stretch his entire body out, now his knees stuck out of the water like mountains making waves, the excess water inching its way up and out of the overflow drain.

He didn't wash. He just sat there letting the knots slowly expand and untangle. The thunder and lightning had passed, but the rain hitting the window provided a soothing, secure sound.

"J?" said Jess. "You haven't drowned up in there, have you?"

"No, I'm still alive."

"Good, 'cause I'd hate to have to dress you before the ambulance came. Nobody should be found naked."

"I appreciate that, Aunt Jess."

"And I hope you not sittin' on the side of the tub makin' those splashin' noises like you used to. If you don't want to take a bath, it's fine. You grown now, but don't waste the water."

When Jeremy came out of the bathroom, Jess was sitting in her mother's rocker. The throne had been passed down to its rightful heir.

"I got somethin' in the kitchen for you," she said, rocking to the rhythm of the rain. He went into the kitchen to find a tin of cornbread. He put it on a plate and returned to her side.

"The table stands are still in the dining room if you want one."

"No, I think I can handle it. It'll be gone in the time it would take to walk in there."

"You always loved yaself some cornbread. Ma Dear loved to see you eatin' it. She loved some you."

"Yeah, I think about her a lot," he said, adding, "Of course, I think of you, too."

"I know you do. But it's easier to think of those that aren't here anymore. But you were her, Patience. Everytime I add another patch to this rocker, I think of her. Sometimes when I'm sittin' here by myself, I talk to her. I call out to her or think I hear her. Guess I'm gettin' senile."

"I don't think that's true. Sounds normal to me. It's almost a prerequisite for living in New York."

"Well, I just know when my number is called, this rocker is gonna belong to you," she said to the air, not even looking at him. "It's in my papers."

"Don't talk like that."

"Talkin' should happen when you able to, not after. If I've learned anthing in life, I've learned that."

The two of them sat there, not speaking and barely breathing, listening to the words they still heard from voices that could no longer speak.

chapter 6

୨୦୨

In the summers of my youth, Mama B would pull her rocker out onto the back porch and I'd sit next to her as we cracked pecans or shelled peas until our palms became darker than the skin that topped them. We'd sit out there, both peering out at the world from behind a screen. But what was out there wasn't nearly as informative as what cracking and shelling brought forth from her.

"Smell like rain," she said, not looking up from an almost-full pan of black eyes staring at her.

"I don't much care for rain. You can't ever go anywhere if it rains."

"If you really need to go, the rain can't stop you. We could use a good rain. Do my azaleas good. Yes, a good rain would be a blessin'."

And the dust would dance, then settle as the rain began.

We'd sit on that porch, mist filling the tiny holes in the screen, forming configurations that transformed instead of disfiguring our view of the world, a world that she knew was in reach.

Our voices filled the space as the rain orchestrated, neither sound drowning the other. Our operas would be of history and how one can learn from the past rather than be saddened by it. I heard of slavery and saw old black-and-white photographs, the kind where the people were alive and present, and like the photographs, fragile to the touch. Those within the frame looked at you as though they wanted to tell you something, whisper tales in your ear, evolving into more than burned images. Though the figures were immaculately adorned, they wore no smiles.

"Those were some rough times, but to say they were bad ain't just," Mama B said to me once when we were talking about the old days, as opposed to the "new times" in which I live now. "To look back and dog ya past is to say you don't like where or who you are now. I'm where I am due to those times and thankful for it. Look what this family has done. I don't think I'm doin' too bad. How 'bout you?"

"No, ma'am," I said.

"'Course, it don't make it right, those times. But that's then, this is now."

Though Mama B's parents were the children of slavery, she had long moved away from those realms. But though slavery had long since lapsed, it never ventured too far from her consciousness.

She had become influential in the community, black and otherwise. Papa Bishop started Bishop's Funeral Home. "'Funeral Homes are good business,' he'd say. 'People always dyin'.'"

When he did what people always do, the business was sold as neither of his daughters wanted to run it and my father wasn't old enough to take an interest.

"Mama B?"

"Yes, Patience."

"How much does it take to be rich?" I asked, as I'd heard it said in the neighborhood that we were rich.

"As much as you have, Patience."

That said, she told me that it was disrespectful and in bad taste to speak of money in front of company. I didn't know if we were rich or not, but I knew I was provided for—yet I somehow felt without.

Aunt Jess was given the rental house when Papa Bishop died and she took charge of it. As far as I could tell, it was Mama B's job to sit in her rocker, smoking her cigarettes and taking her rest.

Thinking on it now, I probably helped kill her, being that one of my chores was going to Scalia's to get her cigarettes. Viceroy 100's. Mama B had been smoking for as long as I could remember and, I'm sure, long before that. I had the good fortune to miss out on the days of snuff, though I had heard tell of it.

"Never deny a person a l'il pleasure," she'd tell me when I'd say how bad they were for her. "We're only in this world for a short time. If you live to be a hundred, that's still no time compared to that pecan tree out there. It was here when I moved here and it'll still be growin' long after we're both gone. So why don't you just take this fifty cents and go get me a pack of cigarettes?"

And I would.

Mama B never opened the pack from what most people know as the top, the easy way. She would always turn the package to the opposite end, the part where the tobacco was exposed, so the filter was never touched before it reached her mouth. With the little pocketknife that she kept in her housedress's front pocket, she'd cut an opening in the bottom of the pack, which thereby become its top. The opening was just at the corner so that the cigarettes could come out freely, yet the wrapping was still intact to protect the fresh, huddled ones that remained.

"Germs'll kill you faster than cigarettes," she'd tell me. "How you handle something can make it more enjoyable. Always take ya time. One less thing to concern yaself with. Just remember, the lazy man works the hardest."

I never saw her use a lighter. I don't think she ever had one. She always used the all-purpose utility wooden matches, the ones that came in the big kitchen-size box. If

*you couldn't light it with those matches, then it couldn't
be lit. I saw them light space heaters in winter, old newspa-
pers at the end of a fishing pole to burn the worm webs on
that old pecan tree in summer, and practically every day,
the pilot on the oven to bake those two pans of cornbread.
If those matches were capable of doing things such as those,
then they could most certainly light a Viceroy 100.*

*"When they get to be seventy-five cents, I'm quittin',"
she'd say, sittin' in her rocker next to the chifforobe that
was its permanent neighbor.*

*The rocker was as complex as she. If you were to peel
each layer of fabric from the seat cushion of that old rocking
chair, each could tell a story so solemn it would still the
chair. It wasn't a rocking chair for an old woman, but a
rocking chair that was old.*

*It was from that chair that I learned about life, love, and
loss. Like a endless scrapbook, knowledge swayed from the
confines of that chair, more knowledge than from the education I was told I was fortunate enough to be getting from
a school I could attend because "we've come so far." Mama
B's education came from life, not school, making her the
best teacher possible.*

*Though the rocker cushion had many patches, some of
which I saw added over those years, one patch was never
covered. It was a denim patch, smaller than the others.
Most of the patches were the same square shape, but this
one was more like a rectangle and was placed on the front
curve of the seat, going from the top to underneath. Denim
was rare, so it stood out, as most of the other pieces were
colored fabrics of some sort.*

*I once asked her about the patch, as I had heard the
stories of most of the visible swatches, but she always seemed
to skip over that one. Mama B, Aunt Jess, and I were sitting
there one evening, the cold of night only something we knew*

was outside the window, laughing about the patches and the stories that went along with them.

I broached the subject. "And this one here? What's that one for?"

Mama B froze, and Aunt Jess also seemed to stiffen, as though the cool breeze had found a crack in the window. I still had a smile on my face, certain that another fantastic tale was to follow. It didn't. Mama B rose and walked as heavily from the room as if the weight of the question had replaced the light thread that hemmed her housedress. Her head was down as if in prayer, chin ducked deep near her sternal notch.

"You awright, Ma Dear?" asked Aunt Jess, but Mama B didn't return with a reply.

Aunt Jess informed me that that particular patch was from my grandfather, and then she went to check on Mama B. I never mentioned that piece of denim again, nor did they.

That chair was the centerpiece of our family. Everyone wanted to sit in it, and often did, but no one needed to be told to relinquish the seat upon Mama B's entrance into the room.

Sitting opposite her in that rocker after many years of being trounced at checkers, I learned to hold my own with her in a game. She knew that game better than Mr. Pressman himself and would often beat me with only one checker remaining with which to defend herself.

"You can't be greedy," she'd say with a laugh after capturing my last checker. "You thought you had me beat 'cause I only had one checker left, but it only takes one to make a king, and the reason it becomes a king is 'cause he watches what's goin' on. He plans, see? You can't go 'round just takin' a checker 'cause it's there. That's greedy. You gotta watch. You always gotta watch for the next move."

Then she'd laugh again while resetting the board, in order to beat me yet again.

"I can't beat you."

"Don't say that. Don't ever let me hear you say can't. I can *beat the hell outta* I can't, *and* couldn't *couldn't give a damn.*"

"Probably," we'd say together.

When the final game of that day had been played, she'd take the checkers away. She kept them in an old matchbox. The box was kept in a drawer of the chifforobe, near the rocker.

In the mornings, which began much earlier for her, I'd get up from the bed that my father had slept in and she would ask how I'd rested. I was young and had no ailments that could so easily be remedied, so I'd always say, "Fine."

Our day would begin with breakfast, sometimes biscuits, sometimes hotcakes, sometimes eggs mixed with grits, but it was always served along with the "Come on down!" cry of The Price Is Right.

She was by far the reigning living-room champion of the game, and I was certain she could beat anyone who had ever been flown out to California.

Mama B hadn't been in a store in "God knows how long," but she knew who was the nearest bidder and how hard to spin the Big Wheel in order to get a dollar, and no matter what, to always take the second Showcase. Many a day, I had fantasized about our going to Studio City, just the two of us. That's the good thing about being a child; the world appears accessible, for the beginner's mind always proves more open than that of the expert.

We never made the trip. Never made a bid. Never got to spin the wheel. After a while, I didn't want to go to California, for I knew that that was where my father lived.

Sitting next to Jess made Jeremy think of all the questions he'd never be able to ask his father.

There were so many unanswered questions Jeremy wanted to pose, but even he didn't have the courage to take what the answers might present. He had picked up some knowledge here and there over the years and could try pressing the shapes together, but it was always frustrating to come to the end of the puzzle to find that those that had handled it before had lost some of the pieces, leaving him annoyed at others as well as at himself.

He walked through the house again, remembering how he had sat on the washing machine and had the ride of his life. He was never asked to do it, but every time the spin cycle began, he'd hop on just for the sake of it and his weight was just enough to curb the sound the machine made because it wasn't leveled. He remembered the teacakes and the pies and the cakes and the jam tarts that filled the dining room on holidays and the tomatoes in the window. These thoughts presented no woe.

"I'd better be heading over to Paul's," said Jeremy.

"Okay. Get that key off the hook there. I know how you young people get when y'all get together. You can stay in the front room if you like or back here in your old room."

Jess had moved to Mama B's room. Everything had shifted to the next.

"All right," he said. "I won't be too late."

"You do need to call Carol in the morning."

"I know. I will."

Jeremy went out to the car. He sat in the driver's seat for a moment before backing the car out. Driving had been one of his favorite pastimes. He'd get in Betsy and drive on I-20, the radio his only companion, playing another somebody-done-somebody-wrong song. Green mile marker after green mile marker. He'd get to the Crossing Pointe exit, then turn around, hating to return home, for home was no longer home at that point. Right after Mama B's funeral, his father had bought a house in Elsewhere, and Jeremy went to live with him—and his new family.

When he got across town, he drove up Lakeshore Drive, which runs along Bayou Natchitochas, one of the most exclusive neighborhoods in the Ark-La-Miss. Most of its tenants had earned or inherited prefixes or suffixes of some sort that dangled resplendently from their names.

He pulled the Oldsmobile in front of a rather large Colonial. Other cars were there as well. He got out and closed the door but had no intention of going in. He just leaned on the car door for a moment, looking at the structure that was his father's house.

3404 Lakeshore Drive.

He took notice that one of the cars was over the curb and its tire sunk into the grass. This stood out, for he knew had his father been there, he would have hated that car and its driver for being so inconsiderate of his lawn. How he had cared about that Johnson grass.

When fried chicken was mysteriously delivered to the house or the tire burns decorated the front lawn or the brick-encased postbox was smashed to pieces, he never had a bad word to say. He never explained what those things meant and Jeremy had been too naive to know. He hadn't realized that being the first to do anything is always difficult.

The front door opened; some visitors were coming out. Jeremy got back in the car and drove on. He could have walked, but this wasn't New York. Here, walking was something that people did from the house to the car or around a track for exercise, but never on the street, regardless of how short the distance. He didn't have to drive far—a mere five houses—for Paul lived at 3414 Lakeshore Drive. He pulled into the circular driveway and the outdoor lights automatically illuminated the lawn. He sat there a minute, wondering what would be awaiting him. He got out and walked through the garage, and just as he was about to ring the bell, Paul opened the door.

"Finally. I thought you were gonna flake."

"Nah, I just wanted to hang with Aunt Jess for a while," said Jeremy, going into the house and then giving Paul a soul hug—shoulders embraced, bodies apart, and two firm taps to the back with a fist.

"Some people are gonna pop by in a l'il. So what's ya poison?" asked Paul, heading to the wet bar.

"Gin and tonic."

"Ah. Good ol' G&T. Mother's milk. I guess we've come a long way from high school and gettin' trashed on Barcardi and Coke and Jägermeister."

"Yeah, I guess we've grown up a bit. How are Samantha and Daniel?"

"They're cool. In Florida. Scoping out for a crib down there."

Paul had always called his parents by their first names, as did all his friends. His parents had once lived in the house that Paul and Beth now called home. Daley House. The houses in the neighborhood were never called mansions; instead, they were always called the So-and-So House, named for their owners, just like their offspring.

Paul's parents were the heirs to a bottling company, and he received a nickel for every bottle or can that came out of the local distribution center.

Paul was the blackest person Jeremy knew, although oddly enough, he was white.

"So, I see you're still the biggest homeboy in town. You look like a walking billboard. They ought to pay you for advertising," said Jeremy, noting the various designer names covering Paul's body.

"Come on, dawg. Why you wanna play a brotha like that?" asked Paul, bringing over the drinks. "I'ma let it slide this time, 'cuz you my ace and I know you pro'bly trippin' due to ya ol' man."

"No, I'm not trippin'. I just thought you would have eased out of it by now. It's me. You know you don't have to do the homeboy thing with me. It's—as you would say—played out. Tired. It's been years since . . ."

Jeremy's voice trailed off and he felt horrible, felt that he was hitting below the belt, taking his feelings out on the wrong person.

"You mean since Pat," said Paul. "Go on and say it. It's cool."

"I'm sorry. It's just . . ."

"I know."

Paul knew a bit about funerals. He was the one who found Patrick, his older brother, lying in his bed. Paul had tried to wake him up, but when he felt him, Patrick was cold to the touch. The sensation had sent a chill through him like the sound of a ringing phone at two A.M.

When the word got back to the kids at school that one of their classmates had died because of a "heart attack," they all mourned the loss of a schoolmate who died of something that was supposed to kill off only old people. The flag was placed at half mast in a semblance of solidarity, and though school wasn't postponed on the day of the funeral, those who skipped class to attend it didn't have it held against them.

Jeremy had come over after the service to see how Paul was doing, to provide whatever comfort he could. That night, they stood in the circular driveway, both silent, unable to find the words that are supposed to be said after a loss. But there was no consolation. They got into Paul's Bronco, and though they didn't move, the radio blasted to the point that the words were unrecognizable, but the message was still decipherable. Anger. Revenge. Rebellion. It was as though the music was what was saving Paul's life; in a sense, he became the music. He became as still as the rattling speakers and neither shed tears nor made a sound, for homeboys don't cry.

The numerous times "nigga" came from Paul's mouth, it didn't offend Jeremy; Paul hadn't had to provide a prologue or epilogue of explanation like so many of his other white friends had. Paul had been the first to accept Jeremy when he moved into the neighborhood, when the criteria for acceptance seemed to be complexion. He had seen in Jeremy something more than pigment, and now it was Jeremy's turn to do the same. He didn't see Paul as a white boy saying *nigger*. He saw Paul, and as he knew so well, grief comes in many exterior forms. How something is said says more than what is said.

Only Jeremy and a few others knew the truth. The Daleys' eldest son had not succumbed to a heart attack. He'd committed suicide. When Paul found him, the bottle of his mother's Seconal was still in his hand. His parents insisted that Pat hadn't taken those pills. It was a heart attack, and that was what the obituary read, what the story was and would be. It must have been true, for it was printed in black and white.

They couldn't have Daley House tainted and tarnished in such a way. They couldn't have something like this say that they weren't the perfect portfolio family. Paul had to play along with the charade. Everyone considered his embracing "blackness" to be a phase due to the loss of his older brother. Some get tattoos or dye their hair orange, but Paul chose another route. The loss he could have handled, in time, for suffering dwindles. It was the lie, lurking like a condor overhead, that made him long to be something else, anything else, that his parents weren't. His new persona became a constant reminder to his parents of the lie; it was a mirror with an ever-present reflection. Their dark secret remained visible.

Paul would pop in and out of the lingo when he was around close friends, but it was something he held onto like crutches in the attic, stored there after the cast had been removed, the break mended. His posturing became his inheritance, providing some sense of security, if not comfort.

Jeremy had always kept Paul's secret, and on occasion, they would speak of it. Paul would drop his diatribe and just be. His parents knew that Jeremy knew. It showed in the way that they diverted their eyes from his. Someone outside the family knew that life wasn't as it seemed and saw them naked. Life wasn't a fairy tale, but Jeremy had known that long before Paul had confided in him.

"It would have been his birthday last week," said Paul, dropping the slang, dropping the manner, dropping his guard. "It still hurts. Every time I walk by his room, I remember it. When they gave the place to me and Beth as a wedding present, it seemed like the perfect gift. But all they really wanted was to rid themselves of the memory. It's been eating them alive. You can still see it."

Jeremy was resoundingly speechless and the room became a vacuum, yet not because of the enormity of Daley House. He didn't know what to say. Although around Paul, he didn't feel that he had to speak, the silence brought back the shield of black.

"But it's all good," said Paul. "It's all good. Here's to ya, Mr. Photographer."

"Fuck you."

"Well, I don't think Beth would be down with that, but it's early yet. Let me see how I feel after a few more drinks." The two started laughing and old times and quips filled their hearts like old tunes forgotten and rediscovered.

"Hey, you," said Beth, wobbling into the room. "Look at you. You haven't changed one bit."

"No, look at you," said Jeremy, standing to hug her. "You're about ready to pop."

Beth was pregnant, but the only sign of that was around her belly. She had always been thin and remained so, even in pregnancy. Beth and Paul were high-school sweethearts and had been together ever since. Jeremy envied that, envied when people knew what it was that they wanted and resolved to make it so. But to truly envy someone is to love them all the more.

"I guess I'm gonna be a godfather soon," said Jeremy, taking on his best mafia-boss voice.

"Any day now," said Beth as Paul helped her slide into her seat.

"Now, Beth, I can see your being a mother, but this one, a father? Boo! Scary!"

"Yo, wassup with that?" asked Paul, taking a rapper's stance, framing his arms around his chest until they rested on his shoulders.

"I'm just pulling your string. You're gonna make a great dad. I can see you and the kid now, both in your Tommy Hilfiger threads, low-ridin' with the sound pumpin'."

"You know him well," said Beth, throwing a pillow over at her husband.

"Listen up. I got a joke for ya," said Paul.

"I can only imagine."

"Just chill and listen. Awright. Three get-it girls are . . ."

" 'Get-it girls?' " asked Jeremy. "What in the world are get-it girls?"

"Are you sure you black? I wonder 'bout you. Beth, J. B. was the only brotha I know who knew all the words to 'Rhinestone Cowboy.' "

Beth laughed and Jeremy covered his face in shame. Yes, at one time, it had been his favorite song, and he'd performed it many times in front of the mirror, certain that Glen Campbell would be proud or jealous. The thought of it now channeled a juggernaut of embarrassment.

Paul went on. "You know . . . get-it girls, like Rolanda Watts in school. Get-it gurrl!" Paul said the phrase moving his head from side to side and snapping his fingers in the air.

Jeremy looked over at Beth with a smirk and she too shook her head from side to side. She was the embarrassed one now.

"Awright, Miss Thang. I got it. But why you wanna bust me about the song? You know that ain't right."

"Nah, it's so wrong it's right. Awright. From the beginnin'. Take two," said Paul, smacking his hands together like a director. "Dese three get-it girls were walkin' 'round the way and it was hella hot outside. Like Africa hot. I mean mad hot. Hot as hot can get—"

"All right, he gets it," said Beth. "It was hot."

"Thank you, Beth," said Jeremy.

" 'Nough from the peanut gallery, please. I'm tellin' this story. Awright, so it's hot, we've 'stablished that. They decide to get a soft drink. So they go into the 7-Eleven. The first get-it girl says, 'Um. I think I'ma have me a 7UP, 'cuz my man is seven and he's always up.' And the other two girls said, 'Get it, girl.' "

Jeremy grinned a bit, more at Paul's telling of it than at the joke itself. Beth just shook her head, rubbing her belly, but the look on her face said that she loved Paul and his carrying on was a part of loving him.

"So the second girl says, 'Oh, I think I'ma have me a Mountain Dew, 'cuz when I say 'Mount,' he do.' The other girls said . . ."

"Get it, girl!" screamed Beth and Jeremy.

"And you know it. So the last girl is walkin' 'round the store and she says, 'I think I'm gonna have some Jack Daniel's.' The other girls look at her and say, '*Gurrrl!* That ain't no soft drink, that's hard liquor!' And the third girl says, 'Dat's my Leroy.' "

The mouthful of gin and tonic came shooting out of Jeremy's mouth. Without as much as a beat, Paul took Jeremy's glass to refill it, repeating the punch line over and over again for the full effect. "Dat's my Leroy."

"And this is to be the father of your child?" said Jeremy to Beth.

"Frightening, isn't it?"

"You are a fool," said Jeremy. "Where in hell did you hear that, or did I just answer my own question?"

"Oh, I work with foster kids in this program that I started up. You'd be surprised what goes on in the minds of kids. But don't try to front like you didn't like that joke. You'll be tellin' it all over New York the second you get back."

"You know me well."

Paul handed Jeremy his drink. "A topper." Though Paul had homebody appeal down, he knew how to make a proper drink, and it never came in a forty-ounce bottle. A fifth, if anything. Some things weren't to be sacrificed. Paul's family waited for five o'clock like most people waited for payday. In Daley House, 5:00 P.M. meant it was cocktail time. If you happened to be there around four-thirty, you would see Samantha, wringing her hands, getting antsy, even though it was certain that she had had several glasses of white wine at lunch just hours earlier.

"Still making them strong, I see."

"As good ol' Samantha would say, 'It's *gin* and tonic, dear. Not *tonic* and gin.' "

"I'd better take it easy. It's different getting drunk in New York—you can just jump in a cab."

"Well, you're at home now," said Paul, hoisting his glass in the air as though giving a toast, "and down here, we learn how to drink and drive at the DMV. Hell, that's part of the test."

Though it was sad, laughter surrounded the truth of the statement. Death, more than anything, had brought Paul and Jeremy back together. They could not speak to or see each other for years and still be able to pick right back up as though they had spoken just hours earlier.

"Well, this get-it girl is pooped, so if ya'll will excuse me . . ." said Beth, breaking the silence. She kissed Jeremy on one cheek as her hand caressed the other. "Good to see you, Jeremy. I hope I'll see you again before you head back up to the Yankees, but this baby keeps me at the ready for the dash to the hospital. This one over here wanted me to give birth in the pool out back. I love him, but he's not all there, you know?"

"I know," said Jeremy, playing along, feigning sympathy.

"Leroy!" she said, taking a terse tone. "I'll deal with you later."

"Promises, promises," yelled Paul as Beth left the room.

"You two couldn't be more perfect for each other."

"Yeah, she puts up with a lot, but it's all good. I don't know what I'd do without her."

"I'm glad," said Jeremy, looking down at his glass. He pulled it up to his mouth, but the stubborn last cube refused to fall.

"What about you? Any lovin' goin' on?"

"Nah. I've been out on a few dates. But I've never been any good at that. Anyway, I don't really have the time right now."

"Another phrase for *scared.*"

"Thanks, Oprah. Can we please just change the subject?"

"Cool. No prob," said Paul, grabbing Jeremy's glass. "Don't you think Charles would have liked that joke?"

"Yeah, I'm sure he would have. He knew a few get-it girls in his day."

"I'll never forget when he came to school to bring you those pants. He stormed in there like he had built the place."

"Yeah, there I was, sitting in my PE suit with my wet, funky pants in that bag. Mrs. Walker says, 'Are you Jeremy's father?' And Charles says . . ."

"No, dahlin', I'm the wicked witch of the South and about to drop a house down on this place," the two said together and burst into laughter. And it felt good to laugh—now.

Jeremy started Bonaparte High in 1983. It was there that he met Paul. Though in certain social circles, private schools were looked upon as taboo, Bonaparte was a public school with private-school appeal, for the lines on the map detailed which neighborhoods could attend, lines that always slanted in its favor. The "good neighborhood" kids drove their BMWs and Mercedeses and Jeep Cherokees and vintage cars, while the "less desirable neighborhood" kids were bused in. Bonaparte High was to become the gumbo that fed civilization with future leaders and athletes. It was thought to be quite clear as to which sector was to accomplish what.

Jeremy found himself in the middle of the madness and was not well liked at first. He was again accused of sounding white, or "country," as it was commonly referred to. The black kids didn't like him because they believed he was trying to be white, and the white kids didn't like him for the same reason. It was the one thing the two groups agreed upon, making familiar and unfamiliar one and the same.

To assure them otherwise, Jeremy began to slack off in his studies, to show them that he was indeed black. It seemed to him at the start that a smart black boy was as recognizable as the dead language of Latin that Bonaparte taught to its future doctors and lawyers. But though his marks fell to failure, there was no convincing anyone and the only one who suffered for his efforts—or the lack thereof—was him.

Once, after physical education class, he had returned from yet another unsuccessful attempt at climbing the rope in gym to find that his pants had been placed in one of the porcelain gods that lined the far wall of the locker room. Without hearing a word of dissent, he went to the office to call home. Jess had answered but was about to take Mama B to the doctor's office, so she said she would call Charles at work and have him bring over a new pair.

Moments later, a call came from Charles, who said he was on the way. He asked for Jeremy's pants size, for he was going to stop by Kincaid's on the way and buy a new pair. Jeremy was pleased by this. He more than expected it, coming from Charles, for Charles had once told him the best way to dilute a crisis was by going into action quickly and quietly. But what Jeremy didn't expect was what form that would take in this instance.

"Where are the pants now?" asked Charles.

"I told you they're in the toilet."

"Go back in there and get those pants out of that toilet, right now," Charles's voice was direct and it seemed tinged with anger, but not at Jeremy.

"What?"

"I didn't say anything about 'what,' Jeremy Bishop. Go get those pants out of the toilet and have them waiting for me when I get there. Do it now."

Jeremy placed the receiver back on the hook as though it too had been placed in the toilet and was unfit for his hand to touch. He didn't understand this request, yet did as he was told. He went back into the locker room, which was now teenager free, but he could still hear the cackles bouncing off every puberty-ridden locker. With a grimacing face, he pulled out the jeans. The smell seemed to override the pine scent that normally filled the room. He walked through the locker room, leaving a trail of splattered waste, and took them to the shower.

He held the jeans with his fingertips, letting the water from the showerhead run through them. His face remained expressionless while he labored. When the jeans were thoroughly soaked and the water running through them once again became clear, he wrung the pants out before putting them in the plastic bag that lined the garbage can. He believed this was where they should have ended up anyway. A janitor's hands, not his own, should have had to deal with this silliness. But in this case, his hands—in color if nothing else— resembled the janitor's and that was what it all came to.

Forty-five minutes later, Hurricane Charles blew through the halls of Bonaparte, waking generations of dust. He took the bag from Jeremy. It wasn't until later that evening that Charles said, "Some people throw punches; some take them—but you'd better learn how to do both if you want to survive." Jeremy didn't believe Charles to be a fighting man and didn't realize it wasn't fisticuffs he was speaking of.

"He was a cool cat," said Paul, again raising his glass in acknowledgment.

"Yeah," said Jeremy, doing the same. "He was like a father to me."

"I know."

chapter 7

∾ ∽

"It'll be good for J to have a man around," said Aunt Jess.
"And I bet you wouldn't much mind it, either."

"Now, Ma Dear, just let well enough alone. Charles is
just a friend. He needs a place, and now that Miss Claire
has passed on, I need to find a new tenant. Charles is good
people, and he works at Allstate insurance, so he can ensure
that his rent will be on time."

I was in the other room, but I heard this conversation;
only children learn to eavesdrop before they learn to crawl.
This particular conversation had to do with the rental
house. Miss Claire had lived there for what seemed like
forever. She was as old as Mama B and just as spirited.
Her sole occupation was dealing with death. Unlike Mama
B, she had never had stake in a funeral home. No, she just
attended them.

She, along with a few other mourning matrons, would
plan their day on the basis of the obituary page, opting for
those write-ups with a photograph so they could compare
how well the morticians had executed their job of preserving
various patrons for perpetuity. They would go to one funeral—sometimes two—a day and had somehow always
made their way up to mourners' row, without being slighted
in the least by those seated next to them. Whether they
knew the "dearly departed" or not, they would grieve with

vigor. Even the family members would sometimes get confused, often consoling Miss Claire and her friends, uncertain as to whether they were close but distant, out-of-the-woodwork relatives that no one wanted to admit to not knowing.

After the funerals, they would go to the repast for the mourners and would tell stories, throwing in "Lawd, we gonna miss . . ."—adding the name of whoever it was that "went on up to bright glory."

When they gathered back at Miss Claire's, I'd hear about the funeral, what was said, who wore what, who shouldn't have been wearing what, and how the food fared compared to that of the previous day's rites.

Every funeral director in town knew Miss Claire and her "mourning maids," and they were always pleased to see them, for they knew a spectacle was to ensue—a true sign of a successful Baptist funeral. Though I had been to a few funerals, they were mostly Catholic ones for relatives of my elementary and junior high-school mates at St. Pascal's. Catholic funerals were always brief. "We like to get 'em in and out in time for the ballgame. Anything over an hour is just sacrilege," I once overheard a man saying at a Catholic service.

Miss Claire went to Catholic funerals only if Gabriel's trumpet was out of tune, placing Baptist deaths on the decline.

"They just don't know how to have a funeral, them Cath'-lics," she told me. "You're good peoples, but you don't know how to cut loose. And cheap with the food, too. Who in they right mind serve cold cuts at a funeral reception? Just a fancy name for luncheon meat. Just cheap, that's what it is.

"And just think they too near to the right hand of God to cry. Like they just allowed three tears during the whole thang, dabbing they little embroidered hankies ever' now and then. That's just wrong. And I'll tell you something *else: they can clutch them little beads till they hands bleed for all I care, but if you can't shout, why even bother to come?"*

Though Mama B and Aunt Jess were Baptist, I went to Catholic church. Each Sunday, I was picked up for Mass by one of my schoolmates' parents.

"If you're Baptist, how can I be Catholic?" I had asked Mama B while we were getting dressed for our respective services.

"Well, it doesn't really matter what religion you are, Patience. It's what's in your heart. But different people worship in different ways; your way is Catholic."

I took this to mean that once again, I was different, and because of that Mama B was ashamed of me and didn't want to take me to her church.

"Is my father Baptist?"

"Well, he is—at least he was raised Baptist. But I don't know the last time he stepped into a church. He's into 'peace and love' right now. I tried to tell him that that was what church was about."

"I would rather go to church with you."

"But you know what? I think your father wants you to go to Catholic church. You know your mama was Catholic," she said, then screamed, "Jessica Bishop, if you make me late again, we're gonna have words!"

"I'm ready! I'm just pinnin' my hat!" screamed Aunt Jess from her room. I could see her down the hall still in her slip. She looked at me and put her finger to her lips, swearing me to secrecy. A car horn blew; my ride had arrived.

"Good morning, Mrs. Christopher," said Mama B.

"Good morning, Mrs. Bishop. Don't you look lovely, as always."

Mrs. Christopher was Danny's mother. She and Danny came every Sunday to pick me up for Mass and the Confraternity of Christian Doctrine classes that prepared us for

our confirmation, first confession sort of thing. CCD classes
were much like—well, if someone said "Three years of CCD
classes or life in prison," you'd ponder, then take the jail
time.

Danny had been my best friend throughout my tenure at
St. Pascal's. Rather than go on to Bonaparte, he went to
St. Anthony's for high school. Danny had been the most
envied kid in school, for his father was Ronald McDonald.
Not always. Most of the time he was Randall J. Christoper,
attorney at law. But on certain days, he was Ronald Mc-
Donald. I wouldn't have known this, but one day, Ronald
McDonald, orange hair, suit, and all, pulled right up into
our driveway on his motorcycle. I refused to go out to the
yard. I never liked clowns, famous or otherwise. Mama B
explained that it was Danny's father. For a moment, I felt
sorry for Danny, but when the Baker boys saw that Ronald
McDonald was at my house. I could then see the power of
a clown.

"Now, here is some money for the collection and for juice
and doughnuts," said Mama B, handing me a dollar.

"You'd better throw in a quarter more for the second
collection for the starving children of Africa," I whispered,
not wanting Mrs. Christopher to hear.

"Well, when that plate comes 'round, just sit there lookin'
real sadlike, and I'm sure they'll pass it right by you," she
said, giving me a love tap on the behind. "I hope y'all have
a nice service, Mrs. Christopher. We'll start 'round ten but
oughta be out this time Tuesday. But Patie—Jeremy has
the key."

Mrs. Christopher and Mama B laughed in their adult
way, but the joke was lost on me. I'd gotten to the age that
having Mama B call me Patience in front of others wasn't
a good idea. Though she often slipped—"force of habit"—
she did her best to monitor her tongue.

As I climbed in the backseat with Danny, we didn't go

into our usual banter. I had found out that my mother was Catholic. Though I didn't want to be called Patience in public, I still practiced the quality behind the name, and bit by bit, I learned more about my mother.

When Miss Claire died, she had the largest funeral in the collective memory of Elsewhere. She had met a great many people over the years during her tour duty of death, and they all came to that final day of resting for her. Every funeral home in the city pitched in to give her a grand farewell, as she had no family of which to speak.

Her funeral was like a party, and I know if she could have climbed out of her casket, she would have reprimanded the entire congregation: "This just ain't right. What in the devil are funerals comin' to?"

Even her obituary said that she was known as "the woman on mourners' row." The photograph had her in the very dress she had worn to so many funerals before and subsequently the one she wore to her own. No one cried. No one shouted. They just passed around stories about when she was at this or that funeral and how she carried on. It was almost like she wasn't even there. Paupers sat next to businessmen, and every awkward combination became less so.

Death had made everyone stop, if only for a moment. As I marveled at it all, I began to think that perhaps that was what death entailed. Birth makes you move, but death stills you. Everyone had put on hold what they were and who they were because in some odd way this woman who lived across the yard from me had touched them, in her loneliness, even if only as an anecdote. She made them be still. Yet, ultimately, that brake in movement segues into a yield, then back to go, throwing caution to the wind. Moving on doesn't mean forgetting, as I well knew and would come to know several more times.

❖   ❖   ❖

Since Miss Claire's death, the rental house, then tan with maroon trim like the main house, had been vacant for some time. Now it appeared that Aunt Jess had finally found another tenant. A man, no less. I was young, but old enough to know that Aunt Jess evidently had no interest in this man—in that way. She had always said, "Havin' a man you like but ain't married to right under ya nose is the best way to end up in jail."

"Charles is gonna stop by later on today, and you'll get to meet him," said Aunt Jess.

"Well, it's your responsibility. You know I don't care about that house, so it's fine by me," said Mama B. "But I don't want no racket comin' outta there. We have a reputation to uphold in this neighborhood. Lawd knows someone needs to keep it up. It's just not like it used to be 'round here. But maybe it would be good for Patience to have a man 'round."

It was settled. Before I knew it, there was a man around, and he was just across the driveway. I had already set myself up to dislike him. I imagined that he had left his son somewhere, too.

When the moving truck pulled up in front of the little house, I sat on the porch and watched him for what seemed like hours. I didn't know how he was going to fit all of those boxes into the little shotgun house. It took three men to unload his belongings and put them on the front porch, while he directed with the ease of a school-crossing guard. He never touched a box. He just stood there in his three-piece suit, in the noon heat of day, but not one drop of sweat glistened on his brow.

When the men had unloaded everything, money was exchanged and they drove away. He walked around the yard. I watched him as though he was going to steal something. He was about my father's build, at least what I could recall of it. He was a full-grown man. He stopped at the flower bed next to the driveway as though looking for treasure. Then he walked over to one of the smaller beds, which was in an old tire painted white. He put his foot on the tire and untied his shoe only to retie it.

"I'll give you a dollar if you'd be kind enough to give me a hand," he said as if he were talking to the petunias. His back was to me. He never looked my way or let on to noticing me, and I dared not say anything. "Not enough, huh? Inflation being as it is, I suppose. You do drive a hard bargain. How about a dollar fifty, then? That's my final offer."

I had immediately decided that he should be sent to Pineville, where the nearest "crazy farm" was. It was clear to me that he had some imaginary friend, which even I, at the ripe age of nine, knew wasn't acceptable.

"You. There on the porch. I'm talking to you." For the first time, he turned and looked in my direction. I stood up, feeling a bit sheepish, as I had been discovered. "So does the price sound right?" I thought again of Bob Barker, my father, and the toy car.

"Me?" I said. "Sorry, I didn't know you were talking to me."

"Sure I was. Who'd you think I was talking to, this rose-bush over here? A rose is a rose is a rose, but it can't talk."

I moved to the screen door and looked at him but still kept the hook tight in its latch, as now not only did I believe he was crazy, I was convinced of it.

"So is it a deal or not?" Though it was inappropriate to speak of money, owing to that, I knew the importance of it. "Yessir."

"Then come on, then. Don't stand there looking all wise and otherwise. I've got to get this place in order by tonight. I'm having a few people over."

"Yessir."

"And you can can it with the sir bit. I ain't ya daddy and the queen ain't never tapped me on the shoulders."

"Yessir—I mean . . . yes, okay."

"Call me Charles. Not Chuck. I hate it when people get real Peppermint Patty–like and call me Chuck. That's not funny. My name is Charles. That's the name my mama gave me and it's the name I like to be called. Charles."

He said the name as though it had twelve syllables in it, which even for a Southerner was stretching it. But from then on, he was Charles to me.

I unlatched the screen door, deciding that though 'I'm not ya daddy,' he was safe. I went across to the rental house and looked at all the things piled in the yard and onto the porch.

"It looks like more than it is. Looks are always deceiving. I'm a bit of a pack rat—refuse to throw anything away. You just never know when it might come back in. Grab those eight-tracks and records there. That's first priority. Once you get the music happening, things move a bit quicker. If they're too heavy, then let me know and I'll get them. Just get what you can manage. Never take on more than you can handle."

I grabbed the box marked TUNES and though they were heavier than the bags of pecans that we would take to the market to sell, I carried them as though my life depended on it. I wanted to impress Charles and show him I was a man too and worth every cent of that dollar fifty.

I put them in the corner of the front room. I thought my back would fall to the floor with them, but I never gave a hint of it. We did load after load. Records, clothes, furniture. Mama B came out to the yard and greeted him, but she never came over to the house. But I know she would have been pleased with his organizational skills.

Charles had a plan sketched out on a piece of paper as to where everything was to be placed. "This way, I already *know where everything should be. It saves some time. Time is money. Some folks take all year to get their things situated. I don't have time for that. Home is home is home, and there's no place like home. Ask Dorothy."*

I scanned my mental map of the neighborhood and my mental list of all of Aunt Jess's friends, trying to place this Dorothy. But nothing came to mind. I assumed it was his wife or girlfriend and I would meet her in time, and I anticipated asking her about home.

Every now and then, Mama B would peek her head out of the screen door. "Charles? Now, if he's in the way, just send him back on over here."

"Oh, no, ma'am, Mrs. Bishop. He's a great help, this one. Better than my ex-friends who thought they were too good to come help me out. I would have been out here all night and well into tomorrow without him," he said, putting his arm around my shoulders. It felt so relaxed there that it was as though he had placed it there for years and my shoulders knew exactly where his arm would fall.

When the last box was brought in, Aunt Jess brought over some lemonade.

"Now, Jess, how do you get the lemonade pink?" he asked, almost as if flirting with her.

"If you don't know, then I'm shole not the one to tell you. A woman has to keep some secrets to herself," she said, and they grabbed each other's hands. I thought they looked so good together—perhaps Aunt Jess did have romantic eyes for him and was finally going to get married.

We sat on his couch—"Sofa, young man," he said. "Couches are for country people"—and drank the entire pitcher of lemonade. I watched him swallow, his Adam's apple bobbing beyond control. I felt my neck, wondering if my Adam's apple would grow to that size. I watched his every move.

"It's good to have a man around," said Aunt Jess. "Makes a girl feel safe."

"Safe? Every man in town knows you carry a razor blade in your bra."

"Stop tellin' tales," she said, laughing at him, placing her hand over the place where the razor was said to be. I couldn't imagine Aunt Jess wielding a weapon.

"Well, I'm sure I'll get jealous once I see all those men streaming out of the house over there, like pulling a number at the fish market."

"I bet you will," said Aunt Jess. It was all laughs and hugs with them, and I beamed as I sat on the sofa. For the first time in my life, having a man around didn't seem peculiar.

My white T-shirt was sheathed with dunes of brown from where I had wiped my face, yet when I looked at Charles, he was as clean as the moment he had arrived. He had taken off only his jacket, but the vest was still in place, and not even a drop darkened the fabric under his arms.

"So, little helper, what do they call you?"

"Who?"

"You? People? Name? Rhymes with fame. What do people call you?"

I thought of this for a moment. I was certain he was mocking me, and the question came as somewhat of a surprise. Everybody around here knew everybody's business. As my many names passed through my head, I didn't know which one to use.

"Speak up, J," said Aunt Jess.

"J? Awright then," he said. "J it is. Well, J, it's a pleasure to make your acquaintance. Thank you for the help. It is greatly appreciated."

With that, he reached into his vest, revealing a money clip, and handed me two dollars. As I reached in the deepmost crease of my pocket for change, change that I knew to be nonexistent, he chuckled. "Slick, this one. I think you worked hard enough for two dollars. Just keep the change."

"*What do you say, J?*"' said Aunt Jess, like I was a very small child.

"Thank you." I cowered, amazed that he had caught on to my I-don't-seem-to-have-change routine. I had only ever done it on women. The fact that Charles was a man was how I relieved my mind as to the loss of that special conniving skill. I knew he probably had his own tricks as well.

"No," he said, rubbing the top of my head. "Thank you."

"Charles, I can't believe how quick you got this place in order. I was gonna have the painters come in at the end of last week, and before I could, you had them already over here. You the best tenant we've had since my daddy lived here."

"Well, I just have no patience for not being settled. I suppose that's how my mama raised me. If the house was a mess or dishes were left in the sink, you were certain to be stirred up in the middle of the night with an extension cord, and the only thing that it was lighting up was your behind."

"I hear that," said Aunt Jess.

But all I heard was my name—the name that I wanted to tell him but never got to. Though I didn't want Mama B to call me that anymore, I thought it would be all right if Charles knew. Aunt Jess had brought this man into our lives, so it was only proper that he would call me what she did.

"Like I told J here, home is home is home and there's no place like home."

"Ask Dorothy," they both said as they laughed and cuddled again. I kept waiting for this Dorothy to appear, hoping she wouldn't dispel the mood.

"Come on, J. We better let Charles get settled in. If I know him—and I do—he's havin' people over."

"You know it. And Jess, you know you better come back over later."

"*I wouldn't miss it. But you know you can't keep up no racket. My mamma's not gonna have it.*"

"*Discretion is my middle name.*"

"*Just be sure that you use it,*" *said Aunt Jess as we walked to the porch, down the steps, and into the yard. I waited for my party invitation to be extended, but it never came.*

"*See you soon, J, and thanks again.*"

"*You're welcome,*" *I said, and Aunt Jess and I walked back toward home. For the rest of the evening, I was so pleased that Charles Discretion lived right across the yard, a new light on a lonely street.*

chapter 8

∼∽

"I thought some people were coming over," said Jeremy.
"Yeah, they should be . . ."

The doorbell rang as the words were coming out of his mouth.
Paul got up to answer it. Jeremy sat in the den, nursing his fourth
gin and tonic, as the other guest walked in.

Hardly a guest at all, as far as he was concerned. There was
Shane Thibadeaux. Shane had once been a friend when they
were at St. Pascal's together, but once Jeremy got him suspended,
they were less than friendly. Jeremy had been the lone black
student there and so took every opportunity to seek retribution
for some of the injustices that take place in junior high school,
Catholic or otherwise.

Shane had been with a group of students one day who were
egging each other on. One of the boys said, "So Jeremy, I hear
Wayne Williams is in town. You'd better watch it. You might be
next." And they roared with laughter. Shane too chimmed in.

It wasn't that Jeremy was hurt by this, for he had made it
through the *Roots* jokes and being called Kunta Kinte behind his
back, but he knew that this subject was a particularly sore issue
and that he could use it to his advantage. Havoc had been going
on in Atlanta, and people throughout the South felt it. Black
people prayed that the murderer wasn't black and whites prayed
that he wasn't white. Regardless, the crux of the matter was that

black children were being killed, making Jeremy if not a prospect, a victim of circumstance.

Jeremy, without as much as a grunt, had marched to the office, mustering up false tears, the sort certain to gain attention. When he relayed the story to Father Warren, phone calls were made to parents and suspensions handed out as easily as if the miscreants were merely being made to kneel in a corner with noses against the wall for misconduct. Shane had tried to defend himself by saying, "I like Jeremy. He's spent the night at my house."

Indeed Jeremy had spent the night, and he left the first chance he got. Shane was the sort who loved his father but belittled his mother to the point of tears. Jeremy had once heard Shane ask her why she was so stupid. His mother had looked at him as though this wasn't anything out of the ordinary. Calling Mama B or Jess stupid would have definitely warranted the murder of a black child.

Lisa was with Shane. They had married while still in high school. An "accident." Lisa went away during her junior year and when she returned, they were parents. Lisa's father had developed new technology for the sun bed and she and all her girlfriends each had one in their homes. Unfortunately, it had aged her. She looked ten years older than she was. Her skin had become reptilian; it looked like old, worn alligator boots.

Paul suggested that they all go out to the pool. As they walked through the house, the questions began.

"I love her. She's a hard girl, but she just wants to be loved like anyone else. But when she's good, there's nothing like her. She'll feed you. It's like any relationship, really. Sometimes you have to say, 'Okay, New York, ol' girl. You win today,' " he said to Shane.

"No, it's just like here—just higher and more expensive. In all honesty, New York is just a lot of Elsewheres divided into sections on an island. There, it's just called the Upper West Side or Gramercy or Harlem. People rarely leave their own version of Elsewhere," he said to Lisa.

"I know a few, but just because you take their picture doesn't mean that you're their friend. It's just a job. I take their picture; then they're off to the next thing," he said to both of them.

"Hell, get off the man. Damn," said Paul, finally stopping the barrage of questions swarming over the aqua water like mosquitoes hoping to pierce a new surface.

"Sorry," said Shane. "It's just everything down here is always the same with nothing much to do, and you're up there, while we have to watch it on TV."

"No problem. I'm used to it. It's a great place to live—or at least work. But Elsewhere is what got me there. It prepared me. It's a love-hate relationship in its most realized form."

Jeremy stared directly into Shane's eyes as he said it, until Shane looked away. Jeremy was reminded that it was all right to move on, all right to remain, but impossible to explain sufficiently.

"Sorry about your father," said Lisa, almost as though she felt she should interject. Her eyes didn't meet his; rather, they focused on the melting cubes in her drink. Lisa's singsong voice had always annoyed him, for the last word of her sentences always ended with an upward inflection, as if she were asking a question.

"Did you kill him?"

Paul burst out laughing and jumped into the pool, Air Jordans and all. When he surfaced, he still had his glass in his hand. He splashed water up on the side of the pool, wetting everyone and making them shriek. It was as though they were teenagers again—but they weren't.

Jeremy walked back into the house, leaving the others to their familiarity. He didn't have it in him for a reunion. The drinks had brought back the memories that he'd left behind geographically, if not mentally. Yes, he had moved to the big city and managed to sustain himself, but Lisa and Shane had something that he didn't—each other.

He walked down the halls, looking into rooms that he had

often frequented. He opened what used to be Patrick's room. The room's contents had been changed, but the smell was stale, like a summer house that needed to be aired at the beginning of a season.

He closed the door silently, walked down the hall, then knocked.

"Come in," said Beth.

"Hey. Sorry to bother you. It was getting crazy out there and I thought I'd come check on you."

"That's perfectly fine," she said, motioning for Jeremy to sit on the bed next to her. "Has he jumped in the pool yet?"

"Of course. It wouldn't be a night at Daley House if he didn't."

"It's his way of breaking the silence."

"I know."

"Sorry for being so antisocial, but I'd just sit out there and moan like a beached whale. I basically just rock back and forth, trying to calm the baby."

"I used to do that."

"Right. Unless I'm missing something, when was the last time you were pregnant?"

"No," Jeremy said, then laughed. "When I was a kid, I used to rock. It made me feel comfortable."

They sat there for a moment, just looking at each other. Where had the time gone?

"So, how's that character out there treating you?"

"He's fine. I think he's nervous about the baby, though. He keeps saying he doesn't know if he's ready. I don't think it's anything you can be ready for. You just do the best you can with what you're given. But he's fine. You should see him with the kids he works with. They love him. It's like we already have a house full of children. He brings them over and lets them swim and he barbecues. But in his heart, he's just as young as they are, and when they leave, you see he misses them, misses the company. I had to hide his high-school ring to get him to stop wearing it."

"But he's grown up a lot. I can tell."

"We all have."

"You're still looking beautiful."

"Yeah, if you like cows. Elsie doesn't have anything on me."

"Stop it. You've always been beautiful, in and out. Though I did worry about you for a while during that bow-in-the-hair-and-ruffled-blouse stage."

"Oh, I hate you," she said jokingly as he ran his hand over her belly.

"I'm really happy for the two of you. Paul definitely needed someone like you."

"We're both lucky. When are you gonna get involved? You know, you can take all the pictures in the world and you can love them, but can they love you back?"

"No. I guess not," he said, letting himself fall back onto the bed. "I can't really say it's that different up there, but it's difficult to meet someone who's ready to say enough is enough. No—I mean . . . people always say that, but they never say it in the way I want them to. It's just that I want to be enough, but . . ."

The words weren't forming for him. She grabbed his arm in a way that told him she understood.

"He worries about you."

"Does he?"

"He can't help it. After Pat, you became like a brother to him. He wants to make sure you're happy."

"I am."

"Are you?"

"Hey, you tryin' to sleep with my wife again?" said Paul, standing in the door with his wet clothes stuck to him like a second skin. "Hell, the least you could do is let me watch."

"Beth, I think it's time we tell him."

"No, Jeremy," she said, "we can't."

"No, Beth. We must. It's gone too far. Paul, Beth and I have been having an affair. Now, I know you think you're black, but boy, are you in for a huge surprise when you see this baby."

Paul took a running start before diving into the bed with them. And in that moment, yes, Jeremy was happy. Enough was actually enough.

"Thanks for having me over," said Jeremy as they walked out to the car. The clouds had long passed, letting the crickets sing and the stars shine. The Big Dipper seemed near enough to sip from. So many stars, but what he wanted was a bite out of the moon.

"Ain't no thang," said Paul. "Sorry about the flashback down memory lane. Shane's always doin' that to me too, and I still live here. He's always sayin', 'You remember the time . . . ?' And I just wanna say, 'Yeah, Shane. I remember—and why do I remember? 'Cuz I was there, and every time I see you, you always bring up the same damn story.' But I let him talk, you know? It seems those times are really all we had. No responsibilities, just living. I just thought you might be into some company. That's all."

"You and Beth would have been enough."

"Thanks, man, but you know, big celebrity and all . . ."

"Just chill it with that—it's *me*. I snap some photos; it's no big deal. Hell, you're married, about to have a kid, and working with foster kids. That makes a difference."

They stood out by the car, and Jeremy scanned the houses and the cars and the manicured lawns. How beautiful they all looked from outside in all their ivory-tower splendor, posing as a Southern Monopoly board, with snake eyes waiting in the night.

"Sometimes it's hard for me to live in this house. Maybe I should have sold it and moved out like my folks did. I just couldn't. I didn't want to run away."

"Yeah, but we had some great times here."

"Yeah. We did. It's all good. And bad."

"It bes that way sometime," said Jeremy as he shoved Paul, and they both began rapping. "I be, he be, she be, we be. If you didn't learn it, I guess ya gotta freebie."

"Ya know, you really oughta chill on the black thing," said Jeremy. "Though you wanna be a brotha, you ain't."

"What's a brotha anyway? We're all brothas when we're six feet under, and then it's too late."

"I hear ya."

"You know the peculiar—"

"Whoa. 'Peculiar?' Now that's a homeboy word for you," said Jeremy, attempting to lighten the tone.

"Shut up," said Paul, like he was fifteen again and they were standing in the driveway. That was where they had always confided in each other, because it was a place outside the houses, where the air flowed freely rather than being encased. "What I was saying before I was so rudely interrupted is: what's peculiar is I can't remember the day he died. I mean, I can remember the day, but not the date.

"Samantha and Dan, they remember the date and always leave town around that time. But they're always leaving. . . . But what I remember is his birthday. It's like I think about how old he would have been, rather than the day that was turned into 'would.' You know?"

"Sure I do."

Their conversation ended with a hug—not a soul hug this time, but a real hug, a hug for the soul.

Though they had shared many things, Jeremy had never told Paul that his own mother had died on Jeremy's birthday; he never told him why he never wanted a celebration and how any attempt at a surprise party would lead to the end of their friendship. He never told him, but he was certain Paul knew. Though he didn't mind that, he couldn't bring himself to speak of it.

I've never much cared for birthdays. While most children would broadcast the days of their birth like a commercial, hoping to generate interest, I remained forever silent. I would listen to them as they reeled off the list of what they would get. Perhaps it was the game Operation, or the grand prize of birthday gifts, a new bicycle.

No matter, they were happy when it was their birthday and they were sure to get something. For me, birthdays were nothing more than a reminder of what I'd lost.

I've only had one birthday party ever. One proved plenty. But Mama B and Aunt Jess were obsessed and felt the end of my first decade did warrant a party. They were at least decent enough not to attempt to surprise me with it.

"Patience, I think it would be nice if we had a little party for you this year," said Mama B, with a smile that a parent might shine on you right before a trip to the doctor. Its purpose is to soothe, but ultimately, it was a smile to downplay what the visit would later entail: a prick and prod, a cough resonating from down below.

"That sounds like a lot of trouble," I said, using my best adult voice.

"Well, it's your tenth, and that's a big one. You need to celebrate."

Though the word why bounced through my mind, I agreed to the party. I knew if I was hearing it at this point, then the plans had already been properly put into place, a party manufactured.

My birthday fell on a Saturday that year. So many kids wish for the calendar to be compliant, allowing their birthday to fall on a weekend, for then they could have a party on the actual day, rather than having to say, "Well, my birthday was Wednesday, but we're having the party today." A Friday birthday might even prompt a sleepover after school, with a party following on Saturday. I knew all the party formulas, for I had experienced them all—for others' birthdays.

When Saturday arrived, I woke early—not because of excitement, but to prepare myself. Each hole on the black telephone dial became familiar to my fingers. I'd place my forefinger in the WXY hole, moving it until the winding sound stopped when my finger bumped against the piece of metal that kept it from going full circle. I'd hold it there for a moment, then release it, hearing the same sound, yet in reverse, the spring uncoiling. Sometimes I'd stop the disc in midrotation, placing my palm in the center, covering the white piece of paper that held the number at which I could be reached.

DEF. ABC. DEF. ABC. TUV. WXY. ABC.

When I picked up my palm, the winding dial would continue its journey home. The black print on the white paper would stare up at me, and I would wonder if my father knew the number or its alphabet version as well as I did.

The phone rang just as the holes found their proper place, startling me.

Mama B came out of the kitchen, wiping her hands with the dish towel. Though my body remained transfixed, I turned my head to her as though I had been caught stealing a Hostess Twinkie from the box she kept hidden in the dining room.

"Gwon and pick it up," she said with an encouraging smile.

"Hello? Bishop residence . . . No. This is Jeremy. . . . That's okay. . . . Hold on, please. It's for you."

The smile left Mama B's face as I handed her the phone. I moved away from the black phone, away from the letters and alphabet, away from the holes, away from the possibility of a father's hello.

I often feared the phone. You see, whenever I'd answer, more than likely I'd be mistaken for Mama B or Aunt Jess. There's nothing worse than a little boy's being mistaken for a woman. Each time it occurred, I felt as if I'd taken a blow to the chest.

I could hear in those voices a sense of embarrassment—for me, not for themselves. They would chuckle out their apologies, like one does when a mistake seems so simple. I would constantly hear Mama B or Aunt Jess say, "That's awright. His voice hasn't changed yet. He's still coming into himself," assuring them that this happened all the time and they shouldn't take it personally. They shouldn't, but I sure did. I made

*numerous attempts to lower my voice, but that never helped.
I was still believed to be a woman, just one with congestion.*

*For me, my birthday was a day to be alone. A day away
from the telephone. A day to do as I pleased. A day like most.*

*I left the house. For some reason, the railroad tracks had
become a playground for me. Unlike most railroad tracks in
Southern lore, it didn't separate black from white. It ran down
the middle, which is where I always seemed to fall. Elsewhere
did have its north and south sides, but unlike most cities, the
north and south sides' populations were reversed. North of
Deckard was the black side of town and south was the white.
"There's no place like Elsewhere," I'd often heard it said.*

*Railroad tracks weren't a social statement to me. They were
where I would become an acrobat walking a tightrope, arms
out just so, to balance myself from falling to the ravenous
lions that waited below for the slightest misstep. To show how
brave I was, I'd often leap from one rail to the other. Should
my balance fail me, it was quite all right, for I would rewind
my inner ringmaster and begin again from that very spot; there
can be no failure when there are no witnesses.*

*When the rails no longer appealed to me, I would walk the
planks between them, step by step, which was easily done,
until I got to the bridge. The planks were wider than my shoe
size, yet the open space between each seemed large enough to
consume me.*

*I'd stand there for a moment. It was only about twenty steps
across. It wasn't a river below, merely a glorified ditch of concrete that kept the water in its place. Rather than step one
foot after the other, I would let one foot meet the other on
every plank until I got all the way across. I'd crossed that path
thousands of times, but each time, the thought of my falling
always tightened my groin rather than my heart. When I found
myself on the other side of disbelief, the tightness would release
and I would throw my hands in the air. I'd made it, and the*

crowd would roar with adulation, astounded by my nerves of steel.

I would then descend to the ditch and sit under the bridge, certain I was in a Tolkien novel. This was my secret place, and the crawfish and guppies that inhabited the shallow stream allowed me to share it with them. I'd sit on the slant with my knees bracing my chin, gravity stretching my thighs.

It was here that I began to wander deeper into the world of making-believe. I'd write the poems at home and recite them here for the school of tadpoles that in time would soon be transformed into something more.

The Coloring of Crayonville
by Jeremy Bishop

The lights came on earlier than usual on Crayonhill,
For it was the first day of school in Crayonville.

The young crayons rolled out of their boxes, their points sharp
    as could be.
They all dressed with their names on their jackets for everyone
    to see.

There was White crayon, Yellow crayon, and Blue crayon, too.
And many, many others, but none of them were new.

The crayons got to school and began to draw many beauti-
    ful things.
Then Ms. Gray, the teacher, appeared upon on the scene.

She had another crayon with her. He had a smile upon his
    face.
"This is Black crayon," she said. "He is new to this place."

Again, Black crayon smiled at the others and slowly went to
    his chair.

But none of the others said a word, as if Black crayon was
    not even there.

Black crayon's family was new in town, and he had left all
    of his friends behind.
His mother, Mrs. Black, said, "You'll make new friends, new
    friends in due time."

But it was not that easy—for the other crayons would not ask
    him to play.
So Black crayon sat in his chair for most of that first day.

He went home crying and told his mother, "This school just
    will not do!"
Mrs. Black said, "Well, it is not going to be easy, but it is
    simply up to you.

"You are as special as the others, and in time they will see.
So be a big crayon and hang on a little longer just for me."

So Black crayon wiped his eyes and promised not to cry.
He told his mother, just for her, he would give it another try.

The next day at school was much the same—until the draw-
    ing test.
Each color went up to the crayon board, adding on to the rest.

But Black crayon was the last and would not get up when
    his time came.
He was afraid of messing up the picture and feeling full of
    shame.

Ms. Gray did not try to make him. I guess she understood.
She just looked at the picture on the crayon board and said,
    "This is very good.

"The picture that you have here is really quite fine,
But it would be much, much better if you could outline.

"Each of your colors gets lost, going into the next,
And that cannot be if you want to pass the test."

Outline *was a word most of the class had not heard.*
So Orange crayon, the brightest in the class, said, "I think it
*   is an action verb."*

All the crayons began to mumble. They wanted to make an A.
Their pictures had been good enough before, so why not today?

Their little crayon faces were all sad. They knew what Ms.
*   Gray said was true.*
But it had been that way for so long, they did not know what
*   to do.*

Black crayon knew he could help but was full of fear.
Then he remembered what his mother had said, and it all
*   became so clear.*

"You are as special as the others, and in time they will see.
So be a big crayon and hang on a little longer just for me."

With this in mind, he looked at his classmates and jumped
*   from his seat.*
He began to outline the picture and did not miss a beat.

He outlined Yellow's sun and the sky made by Blue.
But that was only the beginning; there was more outlining
*   to do.*

He outlined Red's barn, White's clouds, and Green's grass.
And when he was finished, he stopped and looked at the class.

He knew he had done his best. There was nothing he could
*   say.*
All that was left to do was wait for the word from Ms. Gray.

Ms. Gray looked at the picture and began to smile.

She said, "This is the best picture I have seen in quite some
    while.

"And because you all worked together on this, and did it as
    a class,
I am pleased to inform you that you all most certainly pass."

Well, the picture made an A, Black crayon made new friends,
And they all worked together, together from that day on in.

After that class, Crayonville became a more beautiful town,
For no longer at a different color did their citizens frown.

On that day's outing, I had actually forgotten about my
birthday, forgotten that there would be a party in my honor.
But when I returned to the yard, there were the balloons strung
along the clothesline hooked between the house and the white-
painted trunk of the pecanless tree. The donkey was nailed to
the trunk, waiting patiently to be poked and pinned by the
kids in our neighborhood as well as those from neighboring
ones, most of whom I couldn't bear—or worse, didn't really
know. But Mama B couldn't invite one without inviting all.
That wouldn't be right, and it would incite mothers to say
words that were less than motherly.

How appropriate all the guests looked in their multicolored
dunce hats. The cake, purchased from Brookshire's supermar-
ket and decorated according to one of the store's many mock-
up designs, was brought out in all its glory. The green translu-
cent icing spelled out HAPPY BIRTHDAY, JEREMY, and the cake
was decorated with a brown face, though the mock-ups at the
bakery counter never had faces like that. I assumed that this
was supposed to be me, but it confused me, for the face was
smiling.

Ten candles and "one to grow on" were ready and lit.
"Happy Birthday" was sung. I was about to make my wish
when someone yelled, "Eww! He peed on himself!"

It was Pookie Baker, up to his old tricks. He'd let his bladder go just as I was about to blow out the candles. This, of course, took the focus away from the party, away from me, away from my wish. Oddly enough, I was thankful for the interruption. It had saved us all from that awkward, veiled silence that always follows the blowing out of the candles.

Pookie threw his hands up to his mouth and began to cry, and I watched Mama B be Mama B. She consoled him, stopping his tears. "Honey, why didn't you ask to use the bathroom?" asked Mama B.

Chester, his brother, said, "He too scared to ask to go in y'all's house." It became clear to me that Pookie would rather pee on himself than go into the house that seemed like an estate compared to the car lot in which he lived.

The rest of the partygoers had left their seats, trying to escape not necessarily the odor but the thought of the act, as though it was the worst thing a child could do, and especially in public, which elevated the level of shame.

Pookie was taken into the house. "Don't cry, honey. We'll get you outta those clothes and wash you up." Mama B led him into the house, the very house that he was afraid of. Aunt Jess wiped down his seat. The two were the perfect damage-control team. All energy went into making Pookie and everyone else feel at ease. Children ran around the yard, but I sat in my seat. I watched the blue wax of the candles trickle down until it huddled neatly, becoming one with the letters on the cake. "That's one to grow on," I said. And though it was too late to blow, I closed my eyes and made a wish, but when I opened them, all the partygoers were still there.

Pookie rejoined the party with a big smile, wearing my clothes. He looked cleaner that I'd ever seen him. He looked at me almost as though he thought I would rip those clothes off his back and tease him for having peed on himself. I didn't. I did hear Heather Jacobs say, "Hey, Pookie. Urinate, but if you'd cut your hair, you'd be a ten." Though I did find her

statement more than clever, I let it blow over without
acknowledgment.

"You want some cake?" I asked. Pookie nodded, and I cut
a big piece as he reclaimed his seat, not in the least concerned
with whether it had been cleaned. I put the slice on one of
the Fat Albert and the Cosby Kids birthday plates, and he
devoured it. The rest of the kids continued scurrying around
the yard, but I sat there watching Pookie in my clothes and
eating my cake, appreciating both more than I ever had.

"Good?" I asked, already knowing the answer, for his face
had become the smiling one that decorated the cake. "Good.
Good."

Mama B smiled at me, pleased by my restraint. "Happy
birthday, Patience," she mouthed to me. Having mirrored her
manner, I soon smiled too.

I looked at the kids scurrying around the yard. They were
having what appeared to be an enthusiastic time. Everyone
seemed to have forgotten that Pookie had peed on himself,
and the cries of Red Rover filled our street. I played as well,
running as fast as I could, lunging toward the arms I suspected
to be the weakest, wanting to break through and keep running
until I was far away.

As it was my birthday, I was selected as team captain. I
had a tendency to pick not always the best people, but the
ones who needed to be chosen. That meant Rory would have
to be my first selection; otherwise, he would certainly have to
endure the humiliation of being the last picked. He lived with
his mother, but his grandmother lived down the street. He had
come over just for the party. His parents were divorced, but
he saw his father on holidays. At Christmas, his father would
pick him up and take him to the mall to see Santa.

I found out there was no Santa when I was about four
years old. In fact, I knew there was no Santa before I knew
who my father was. Aunt Jess told me, which made me dread

Christmas—not because there was no Santa, but because there were adults. Daily during the holiday season, I was subjected to some adult asking me what Santa was bringing or had brought me. I'd look up at Mama B, my eyes saying, Maybe you should break the news to them. *She would explain that I already knew about Santa. It always made them look at me as though I was as deprived as a nest on a leafless branch in winter.*

But Rory believed, and I knew better than to mention what I knew to anyone who truly believed. On one occasion, I accompanied him and his father on an outing to Pecanland Mall—so named to memorialize the trees that died for its benefit—to have his picture taken with Santa. The invitation had been extended to Mama B on my behalf and was accepted. Mama B always wanted to appear inclusive and not protective of me, so I was off to see Santa.

We waited in the believers' line. When Rory's turn came, it was memorable.

"Have you been a good little boy this year?" asked Santa— or at least the man playing Santa.

"Yes, very," said Rory, who was wearing his best been-a-good-little-boy, about-to-take-a-picture-with-Santa outfit, which he had probably picked out and ironed all by himself at the ripe age of six.

"Now, you wouldn't be telling Santa a story, would you? You know, Santa knows when you've been naughty," said the man, closely following the script he had repeated for as many kids as time would permit. He could say the same thing over and over without fear, for his house was surrounded by a picket fence, so none of the other believers could hear what was going on before their turn arrived.

"I know, Santa," said Rory. "But I've been very good."

"Then what would you like me to bring you for Christmas?"

"I would like an Easy Bake Oven."

"No, no!" bellowed Rory's father, standing with me. He was

as embarrassed as if Rory's pronouncement had been made over the public announcement system. "Tell Santa you want a Tonka truck or a football or something like that."

"No!" screamed Rory, topping his father's plea, "I want an Easy Bake Oven!"

And rightly so—they were exceedingly popular at the time.

When the last balloon had been popped and the yard raked of debris, I went into the house. I moped around, my shoes like lead, still waiting for the phone to ring.

"Do you want to call him?" asked Aunt Jess.

"No," I said.

"I'm sure he'll call. There's a time difference, you know."

"I don't care."

"Now, don't say that, J."

"It's true."

"Sorry about today," she said, and finally the refuge came, but not from the telephone. Years of tears had been huddled near my heart for hours, but when they surfaced, their sound woke souls. It was then I realized that tears help, but sobs are the true release.

Later, I heard Mama B on the phone. Though I heard only her end of the conversation, I knew with whom she was speaking. I had never heard Mama B raise her voice; it was not in her character. And though I'm certain she and Aunt Jess had their rows, they never did so in front of me. When she put the receiver down, I could see the steam prints that her grip had left. She reached for her inhaler of bottled breaths and sat in her rocker. She didn't say anything. She sat there, breathing heavily, then slowly began to rock.

"There's the birthday boy," said Charles, getting out of his car. "How was the shindig?"

"Fine," I said, walking in the very yard that hours earlier was filled with lost wishes.

" 'Fine'? Fine is something you say when someone asks you how school was, not something you say about a party."

"It was fine, awright?"

No one has ever taken my fits of outrage seriously. I suppose they seemed so out of context with my usual demeanor that when they did occur, they usually brought laughter from those that truly knew me, pushing me back into complacency.

"Fine, fine. Do you think you could give me a hand with some of these grocery bags?"

"Fine."

"And you need to put that bottom lip in before you trip over it. For a boy your age, you've got the biggest soup coolers below the Mason-Dixon."

I refused to laugh, though Charles was doing his utmost to bring me back to normal, if there was such a thing.

"Just put them in the kitchen," he said, filling my arms with two bags.

"Where else would I put them?" I shot back.

"Awright, smarty pants. I guess you think you're grown now?"

"I guess so," I said as I walked to the kitchen with his two bags. I placed them on the counter. Though I didn't trip over my lips, I did almost fall over the red Schwinn ten-speed bike with a big yellow bow on it.

Mama B and Aunt Jess had joined Charles and they were waiting outside. When I ran out to the front, they screamed, "Happy birthday!"

"That's from the three of us," he said. "A man's gotta get around, and we figured since your other bike was stolen, it was about time that you graduated to a ten-speed."

"Well, J, gwon and bring it out. Take a spin," said Aunt Jess, as though she had been given the bike and anticipated the first ride. I went back into Charles's kitchen and wheeled it out. I boarded it and rode around the block, beaming. Aunt Jess had taught me to ride a bike years ago. I remembered her *being out in the middle of the street, running behind me, holding the seat until I got my bearings. She was the only training wheels I ever had. It must have been a sight to see, Aunt Jess running and Mama B traversing the porch, screaming, "Ride it, Patience! Ride it!"*

The Baker boys had gotten a kick out of it, too.

"Can't no woman learn you how to ride no bike." It was widely known that teaching a child to ride was a man's job, but rules are often changed as necessity dictates. Necessity was why Aunt Jess taught me to ride. The scars that the blacktop gave me were different than the numerous ones present, yet far from visible.

After a quick spin around the block on my new ten-speed, with its big-boy crossbar, I got off and hugged Aunt Jess, Mama B, and Charles. The memory of that day and the party had dissipated, and life was more than fine.

"Is that the phone?" asked Mama B. I looked at her, and all of us perched our ears with the curiosity of dogs. We started to run toward the house, the rings up to three, then four, then five.

"Hello?" I said, grabbing up the phone.

Dial tone.

chapter 9

～❧～

Jeremy returned to the Oldsmobile. The smell of it seemed to intensify as he sat in the seat. How many others had called this car theirs for brief amounts of time? All the scents melded, yet none were distinct to his nostrils. Now, the faint smell of liquor would be his contribution.

He drove out of Paul's circular driveway and the lawn lights again illuminated. Everything seemed full-circle as he again drove past his father's house. The lights were out and the cars had made it back to their respective drives. But there was someone standing in the driveway, cloaked in a shadow of reality. It was Jason, his half brother.

Jeremy took his foot off the accelerator and let the car roll at its own speed, contemplating whether to brake. When the headlights flashed on Jason, he put his cigarette down by his side, cupping it in his hand. He waved at the car, the Southern salute always extended to folks both familiar and unfamiliar—and even when one was unsure of which it might be. Jeremy didn't pull the car into the driveway; he parked it on the street.

"Hey," said Jeremy, opening the door and getting out of the car. He walked toward Jason, his steps as if on just-poured concrete through which his feet were certain to sink.

"Wassup?"

Jeremy passed up his normal reply to that question and said, "Trying to make it. How about you?"

"Same o' same o'. I heard you were comin' down."

Jason's voice was of a low register, the words hiding under his headful of dreads.

"Yeah, got in today. I was gonna stop by earlier, but . . ."

"Whateva," said Jason, finally bringing the cigarette back into sight and taking a drag.

"Can I bum one of those?"

"What? You mean the almighty good son smokes?" He pulled the pack out of his pocket and passed Jeremy a Newport. After he replaced the pack, he pulled out a lighter from the other pocket. The lighter clicked, smelling of naphtha, and he lit the cigarette for Jeremy.

"Nice Zippo," said Jeremy, feeling a slight singe on his brow. "Umph, menthol."

"Beggers can't be choosers."

"I've been trying to quit, but when I drink, I—"

"No! The almighty good son smokes and drinks?"

"Yeah. Do they know you smoke?"

"Carol does, but I can't smoke in the house, and she doesn't really like me to do it in public—you know, appearances and shit. As for the ol' man, I don't think he's too worried about it right now."

"I suppose you're right."

They stood there together, finger-flicking the cigarettes at a furious pace to rid them of ash that couldn't keep up the pace. They inhaled vigorously, the lifetime that separated them as impenetrable as the smoke screen they were creating.

"So. How's BH treating you?" asked Jeremy, grasping for contact. "You're what . . . fifteen, sixteen? What's that—sophomore?"

"You ain't too good at this, are you? I'm sixteen and I'm a junior at Riverwood, not BH. BH is 'too dangerous. It's just not the same as when Jeremy went there,' so they stuck us in Riverwood."

"Private. That's cool. Great school. I knew a few people that—"

"Is there a point to this brotha-to-brotha chat?"

Jason took an aggressive stance to let Jeremy know that he wasn't into chitchat. He stared straight into Jeremy's eyes, wanting to provoke battle rather than banter.

"No. I was just passing by and saw you out here and thought I'd stop, say hello."

"Well, ya oughta fuckin' call Guinness, 'cuz that's the longest hello I've ever heard."

Jeremy took a drag on his cigarette, blowing out the smoke before it could even taint his mouth. He couldn't bear menthol cigarettes. Jason took a long last drag, then threw the butt on the driveway. For lack of something to do, Jeremy stepped on it and picked it up.

"Well, I guess the good son can't leave a speck of visible filth in the yard. Just like the ol' man, aren't you?"

"What's your problem?" asked Jeremy, taking offense at his younger brother's aggression and implications.

"You're my fuckin' problem. The best thing about the ol' man being dead is I don't have to hear your name every fuckin' day. They should have just tattooed *Jeremy* on my ass so they could kiss it every time I walked by."

"Well, at least you had him around—"

"Oh, God," said Jason, caustically. "Not the 'my father abandoned me' sob story. That's played out. Don't you think you're a bit too old for that now?"

"You're a bit young to have that big chip on your shoulder."

"Take a look in the mirror."

"And I suppose this is your car," said Jeremy as they continued topping each other, like the frantic building of a musical score out of control. "I see you didn't refuse that. And I'm sure you have a job to pay for these cigarettes you're smoking. And this little rebel thing you're doing is pathetic. I'm sure that goes over well at Riverwood."

"I'm pathetic?"

"Yeah."

"*I'm* pathetic?"

"Yeah, you."

Their voices continued to rise until a light came on in the driveway, signaling a temporary cease-fire.

"Jason?" It was Carol. "Oh, Jeremy. I didn't know you were out here."

"Hi, Carol," said Jeremy, his tone returning to affable.

"Is everything all right?"

"Yeah, fine," said Jeremy, hoping to disguise the tension. "We're just catching up."

"Listen, man, I don't need you to cover up for me. I've been doin' awright for myself for a while, so save the brotherly love for someone who buys it."

"Jason!" screamed Carol. But his name caught only his back as he walked through the garage and into the house, leaving Carol and Jeremy standing with the reverberations. The exchange had not gone as Jeremy had anticipated. He had just wanted to say hello. He hadn't come to fight and he hadn't wanted to see Carol—not yet.

"He's been upset."

"That's understandable. He's lost his father."

"Oh, no. He was upset long before that. He's a teenager; being upset is his job," said Carol, throwing her hands up into the night, but far from defeated. "He's convinced that the world and its populace are here just to ruin his life. Maybe you could talk to him."

"I think he's made it rather clear that he isn't interested in talking to me."

"He's just lashing out. He talks about you constantly. I just think he feels he has to live up to what you've accomplished, and it doesn't come as easily for him as it did for you."

Jeremy bit his tongue, wanting to scream. He looked at Carol. Her face said she hadn't slept in a while. Her eyes were tired and weary. Here she was in this big house on Lakeshore with two kids. She'd lost her husband and they'd lost their father, yet she still tried to find a way to console someone else.

"I'd better be getting back. It's late. I know Aunt Jess will be waiting up."

"You can stay here if you like. Your room is just like you left it."

"Thanks, Carol, but I think I'll just go back over to Mama B's."

"That's fine," said Carol, yet they both knew that it wasn't. Once again, she had made an offer to him, and once again, he had refused.

"I'll call you in the morning."

"Yes, do. I need to go over some of the details with you."

"Yeah. Try to get some sleep."

"I'll try."

They stood there, waiting for someone to move, to provide a transition that would lead to an exit. The moment lingered into two. Without another word, Jeremy just turned and walked to the car. Carol watched him. As the engine started and his car began to move away, she waved.

My father called Mama B to inform her that he wanted me to meet his new wife and family. I gave him credit for not calling her my new mother—or perhaps he did, but that wasn't how the message was relayed to me.

Even though I wasn't convinced that I had a choice, it was Charles who urged me to take the trip. I went over to his place and talked to him about his parents. He had pictures on his walls of friends and family. He pointed out everyone, telling me who they were and the stories behind the shot. When he showed me his mother, he paused for a moment. He always spoke of his mother, and I always felt that he was fortunate to be able to do so.

"Is she ever gonna come visit you?" I asked.

"No, J. We don't talk much anymore."

"Why? She must be proud of you. You should call. . . ."

"It's not as simple as that."

"Why?"

"We just don't have much to talk about. We don't see eye to eye on certain things."

"But you love her, don't you?"

"Of course I do, J. Sure I do. But sometimes in life, that's just not enough. Things that are a part of you will always be a part of you. But sometimes you have to get away. Have some distance."

Distance.

This sounded to me like something that a man would say and seemed only fitting in the moment. I couldn't imagine having a mother and not being able to speak to her. It was one thing not to talk to my father, but I knew if I had a mother, we were bound to be close.

"My father wants me to visit him in California," I said, changing the subject, though the subject seemed the same.

"Yes, I heard. Sounds like fun."

"Not to me. I don't want to go out there."

"Well, he's trying to reach out to you. At least he's trying."

"Whatever."

"No. Not 'whatever.' It is something. I tell ya what. I've never been to California, so why don't I let you use my camera and you can take pictures of everything. That way, I can see it when you get back. I hereby declare you the official trip photographer."

Charles had found a way to distract me from my ambivalence. He got the camera out of a drawer and placed it around my neck. It felt heavy as it hung there. I thought about lost spirits and the Indians. I thought that if I took enough photos of my father and his new wife, their souls would slowly disappear, and with them, the memories.

Charles sent me back over to Mama B's with my new assignment. I still had doubts, but Mama B insisted that I should meet Carol and my brother and sister. I was almost thirteen and they were three. Because of the age difference, seeing them was all that I could possible do. What more could there be?

I didn't understand why, if my meeting them was that important to him, they couldn't all come see me. Mama B explained that that was because there were four of them and only one of me. She didn't mean it the way I took it.

Carol was an attorney. She met my father in "his chair," as though the good dentist had looked into her mouth and found love. Eight months later, they were married. "It wasn't a big deal, just a few close friends and associates," my father told me. I wasn't invited. It was probably best, for the both of us.

Seven years had passed since the toy car had pulled out of that driveway. I'd spoken to him several times on the phone, but the calls were always brief. In order to pick up where you left off, there has to be something to pick up. Our conversations were usually banal, at best. I knew the answers to his questions as though I had studied for an exam.

"How is school?"

"Fine."

"Are you looking out for your Mama B and Aunt Jess?"

"Yes."

"Well, hang in there."

"Okay."

"Pass me back to your Mama B."

Mama B explained that my trip would be for only five days. "Three, really, 'cause the flight is a long one."

Aunt Jess and Charles got in the Rambler and they accompanied me to the airport. Mama B wasn't feeling well, and I tried to use that as an excuse for not going. "If you go, I'll feel better," she said. For some reason, I couldn't associate going with feeling better.

I got on the airplane and the stewardess showed me to my seat. She complimented me on my camera and asked me if this was going to be my first visit to San Francisco. I told her it was my first flight anywhere from Elsewhere. She said that I shouldn't be nervous and that planes were safer than cars.

I wasn't so much afraid of the flight—I was more concerned with what awaited me on the other end.

The stewardess, "Mandie," smiled way too much, so I took a photo of her, hoping to rid her of it. She rattled on about how I must be a very special "young man," flying in first class. First class had no meaning to me, but it seemed to mean a great deal to my father and everyone else that heard that I was going out to California first class.

What did stick in my mind was the part about "special." I always thought of that word as pertaining to school. We had gifted and talented classes, regular classes, and special education. Though I was in GT, as it was called, I always wanted to be in the special class, for I realized that special meant that they didn't quite know what to label you. No name had yet been given for your particular ability or, as they thought of it, lack thereof. Because of this, a great deal of attention was shown to the kids in the special class. They had to be nurtured and cared for in a manner different from the rest, for not everyone has the capacity to deal with a special child.

As the plane began to taxi, I looked at the spot where Aunt Jess and Charles stood. I could see them waving. I stared at them right up to the point where all I could make out was the place I knew they were standing. Though I was certain they couldn't see me, I waved, too.

After a layover in Dallas, the plane landed in San Francisco a few hours later. The stewardess, still smiling, saw that I was guided off the plane and to the gate. She waited with me, but no one was there to receive me. I was in a haze comparable to the one I'd seen hanging below as I had peered out from my window seat during the descent. Having Mandie at my side made me feel like a child who had wandered from his parents in the mall and was subjected to the humiliation of everyone knowing that he had parents who didn't pay attention to their child. I felt the looks down the nose that were sent my way. But I didn't at all care about how it would make

my father look. It was my embarrassment that concerned me.
I had not strayed.

The courtesy call was left unanswered. I was then taken
down to the baggage claim area. Mandie explained that she
had to leave because she had another connection, but that
"this nice man" would stay with me.

This man was in a suit, and I gathered he worked for the
airport. I answered his questions in a monotone. They seemed
like flash cards used to test my intelligence.

"I'm here to visit my father."

"No, I don't know the number."

"No, I don't know the address."

"No, thank you."

"No, thank you."

"No."

"I'm sorry."

The last reply, though it was mine, seemed to be his thought
as well. I was somewhere far from Elsewhere, and though those
questions would have been an insult to me at home, here they
were just a reminder of how little I truly knew about where I
was and who it was I was there to see.

After an hour of sitting in an uncomfortable plastic seat,
watching people that knew where they were going, I wanted
to take my ticket and get back on the next plane. I remembered
Mama B assuring me, "It's a round trip. You can always come
back early if you want to."

Three airport employees had taken their shift with me, and
still no sign of my father. I tried to imagine what he'd look
like after all these years. Two hours later, it was my turn to
hear my name echoing through the terminal. The employee
handed me the courtesy phone. It was Carol. Her voice was
violin-string tight, evoking stress that seemed out of character
in a place where everyone's philosophy seemed to be "No
problem." I had had plenty of time to notice that people in
California moved as slowly as they did in Elsewhere.

I sat waiting for this woman I'd never seen so she could rescue me from the "in holding" room. When she arrived, she had Jason and Jessica with her. She gave me an uncertain stare, which I didn't return. I had been taught not to associate with strangers.

"Jeremy?" she asked, a question and statement intertwined. The employee walked me over to her. When the proper identification had been presented, revealing our mutual surname, he released me to her care. I wondered what he thought as he watched what I'm sure he assumed was a reunion. No hugs were exchanged, just guarded gazes. She thanked him, opening her Louis Vuitton purse and pulling out some folded bills, which he declined. She thanked him again, again pushing the bills toward him. Again, he declined. He squatted in leap-frog position and told me to have a nice visit. He was smiling.

I was introduced to my brother and sister. "Jason, Jessica, say hello to Jeremy. He's your big brother." No hands were outstretched. Jessica grinned and did a little curtsey, but Jason just stared as he tucked his hands in his OshKosh overalls.

It was in her Volvo station wagon that it was explained to me that my father was supposed to have picked me up but had been "delayed with a patient." She apologized for him. She asked how my flight was. I told her it was fine. But it was a horrible flight—it had landed.

Jason and Jessica sat in the backseat, the latter smiling, the former still staring. Carol and I sat in the front, looking straight ahead. Every now and then, I would put the camera up to my eye just to have something to do, yet the shutter never moved.

The car ride seemed to take as long as the flight. Traffic kept us on a bridge that went on forever, compared to the tiny bridge that separated Elsewhere from West Elsewhere. I think I held my breath the entire time we were on it, certain that we would find ourselves in the water below before we made it to the other side.

I looked at Carol sitting next to me. I could see that she was uncomfortable. I could see her attempting to formulate sentences to say, then resorting to remaining silent. We both made use of peripheral vision. Though I had imagined that she looked like Mrs. Richards, the neighborhood witch, I did have to admit that she was more than attractive. I knew Jason and Jessica were too young to be embarrassed by the fact that their mother was so pretty.

When we finally arrived at the house, I was shown my room. Jason and Jessica followed my every step as though I were a new toy they had discovered, with lots of gadgets to be explored. My room was spacious. I was almost too afraid to touch anything, fearing I would break something and draw unwanted attention to myself. I looked around, thoroughly, but no sign of me existed. I was told my father would be home soon.

"There they are" was what he said when he walked in. Jason and Jessica ran over to him, knowing who he meant, but I can't rightly say I knew that I was included in that they. "What? No hug for your old man?" I walked over and stuck out my hand, proffering a shake, as I had seen other strangers do when introduced to one another. He obliged, saying wasn't I the little gentleman.

He had a present for me. It was a backgammon set. I knew this only because that was what the words on the box said. When I opened it later, I recognized the board. I had seen it on the back of the checkerboard that Mama B and I used. We never played with that side. I had always assumed it was just there for decoration, something to fill the space.

I later heard him arguing with Carol, at Richter scale levels.

"There's no excuse," she said.

"I know. I'm sorry. Get off my back already," he said. "What did you want me to do? I was in the middle of a procedure."

"You don't have to tell me. I work, too. I have clients, but

you've known for weeks that he was coming. Patient or no patient, it's inexcusable. You make time for what it is you want to make time for. You should have seen him. He was sitting there for two hours. How do you think that made him feel?"

"I know. I'm sorry. You're right."

"Well, damn it, I don't want to be right—I want you to not be wrong," said Carol, staring him down. I could see them down below from my perch atop the stairs. He looked at her and her tone changed, her voice lowered, the jury convinced. "This isn't easy. I know. But you've got to make an effort. He's your son. You're suppose to be there for him. This isn't the way I wanted to meet him. He didn't even know what I looked like, and there I am having to introduce myself. That should have been your job."

I pressed my head against the banister as I listened. I was to figure out later that in California, arguing was called "having a discussion," but their discussion sure sounded like arguing to me. I kept eavesdropping for a while more. It ended with my father saying that he had taken the next few days off and he'd make it up to me.

That was what I recall of the first night, but all the nights were the same as far as I was concerned. I wasn't impressed by California, and my lethargy seemed to fit right in. I snapped my photographs at will, thinking of Charles and what he'd find interesting. When I told my father that Charles had given the camera to me, he seemed nonplussed. The backgammon board had not yet been opened.

We drove around the "city by the bay," but all I knew about this place called San Francisco was that Rice-A-Roni was its treat, yet not once was I treated to it. I was told the big bridge that frightened me was the Golden Gate. We went to Chinatown, and I truly thought I was in another world. The whole animals hung in windows, their golden brown exteriors inviting to some, but the thought of consuming them made me squeamish.

Fisherman's Wharf. The streetcar and the hill it straddled. I looked for the woman in the commercial hanging wistfully in the rear, but she wasn't to be found.

Alcatraz. "How would you like to live there?" he asked, but the idea of it didn't seem foreign to me.

His office. He sat me in his amusement park–like up-and-down, back-and-forth chair, then looked into my mouth. When his fingers pulled my cheek aside to give him a better view, I wanted to bite them off. It made me think of the many teeth that I had lost that he never saw, and about how I pretended to believe in the tooth fairy just for the dollar that I would receive and the pleasure that Mama B and Aunt Jess got out of removing the tooth from under my pillow. I don't know why they figured that if I knew there was no Santa, I'd buy the story that some fairy visited and left money, but as long as the dollars came, I was inclined to play along. I was no fool. I remembered those times as he perused my teeth, but I knew he didn't think of them.

The day before I was to return to Elsewhere — home — we all piled into his blue BMW and went to the beach. It was my first visit to the ocean, and I quite enjoyed it. The sound of the water was lulling, and the waves that overlapped as people rode them back into shore made me forget that I was supposed to have a bad time. My resistance ebbed.

I sat on the beach in the swimsuit they purchased for me. I laughed and even assisted in the hunt for shells and pebbles. Elsewhere was nowhere in mind. We all went into the water. Though they seemed to believe it was warm, it still seemed cool to me as I inched my way farther out on tiptoes.

Jason and Jessica had yellow floatees on their arms. They were having the time of their lives — or perhaps they were used to it, and this was life in time to them. I had stood for a *while on the foreshore, trying to adapt to the water and its conquest of my body.*

My father was laughing and splashing water on Carol. They did appear to be what I thought a family should look like. They looked happy, no discussions. Jason and Jessica took turns sitting on his shoulders, afraid of nothing, but when he asked if I wanted a turn, I declined.

As the waves came in, I would jump up, topping them just as I had seen the other beachgoers do. A wave would lift and transport me nearer to the shore, then I'd walk out a little further to repeat the sensation. Perhaps I became too secure in my newfound skill, because I found myself wanting to try bigger waves.

I saw several waves forming into one, coming in toward the shore and mounting in intensity. I jumped up, my timing a beat off, and something grabbed my legs from below and pulled me under. It wasn't a person; it was much larger than that. It was the Pacific, having her way with me. My eyes remained shut, for I had already discovered the irritation salt water could cause, and panic began to set in. I began flailing my arms and punching at her, with little result, hoping to make her unleash me, hoping I would get to breathe again and release the pressure in my ballooned cheeks.

Finally, at her leisure, she obliged, and my head shot up like a beach ball forced below the water and then released. When I surfaced, the sound I made was the scream of life made by one who believed he had been about to die. I was stunned, bobbing yards away from the others. Shock made the water even colder, and the breeze brought chill bumps to my skin. I focused on the picture-perfect family, playful seals shining in the leisure of the day.

Without as much as a word I started back, again finding myself on the ebb. They all remained in the water. I sat on the beach for the rest of the outing, realizing that saying the wrong word or revealing that wrong emotion could easily turn the tide, making castles into what they really were—merely grains of sand waiting for the wave of erosion to take its course. No one had seen me go under, and no one would know that I had.

Carol soon joined me. "Having fun?"

"Yes, thank you."

"Well, we've enjoyed your being here with us. Your father speaks of you often."

I didn't say anything, but I wanted to know what he had to say about me. For some reason, I couldn't hate Carol. I couldn't say I liked her, but I didn't mind her. Jason and Jessica were cute, as kids often are, when cute can still be taken as a compliment.

At the end of the trip, I discovered that I had been brought here because my father and his family were considering moving back to Elsewhere so that he could open a practice there. He said that with Mama B's health as it was, perhaps it was time to be nearer to family. But their move would come too late.

Many things stayed with me from that trip. I had seen my father and his family. I had seen San Francisco. I had been to the ocean and learned to respect her power. And when I returned to Elsewhere, sand accompanied me in my shoes, insinuating themselves into my existence. I took my shoes out to the yard and looked at the grains huddled in the heels, far from their home. I poured them out and watched the crystals bounce on the driveway until they seemed to disappear, for a few grains do not a beach make.

chapter 10

∽⌒∾

When Jeremy pulled away, he thought of his day. How long it had been—and that wasn't due just to the extra hour that the change from Eastern to Central had provided. His morning had begun in New York, and now he was farther away than the miles he'd traveled would indicate. Aside from dealing with various questions concerning his life in the big city, he'd not thought about New York. He hadn't called to check his messages, which was unlike him. Whatever was happening there could wait, for a change. He was actually pleased to be away. But the reason for this journey hit him again, slumping his body behind the steering wheel until he became one with the Oldsmobile's velour seat.

He drove for a while. Jason kept stealing the focus from his thoughts, becoming a vision as strong as Paul's gin and tonics. He saw so much of his younger self in Jason. Cocky, angry, and perhaps—yes, he dared think it—spoiled and self-absorbed.

He kept driving, but it was as though the car was steering itself. Before he knew it, he was parked near the cemetery.

Mama B had been buried in what was then referred to as the "colored cemetery," next to her husband. "A woman should always be married out of her father's house and a wife should always be buried next to her husband. That's all that matters," Jess had said when Jeremy's father and his sister Gladys protested about the condition of the cemetery.

It was true that the surroundings weren't at all heavenly. Yes, the grass was cut on occasion, but the place didn't give anyone much pause, because of its general unkempt look. The only people allowed to find their rest there were those who had already secured plots, which at the time was a grand show of achievement in the colored community. Then and now, the graveyard was a reminder of how far they'd come and exactly who they were, depending on how one chose to look at it. Mama B would be the last in their family to be buried there—progress would separate her from the rest of the family.

The cemetery or its condition had not mattered to Jeremy. He wasn't there to think of community advancement; he was there to pay tribute. He'd seen his father weep at the funeral. He'd wondered if those tears were regret for the loss of life or regret for the life he had run from, leaving others to mind the responsibilities he'd left behind.

As Jeremy sat in the Oldsmobile, the funeral of years ago came back to him. He got out of the car; the gin and tonics made him stumble a bit. He laughed as he walked into the graveyard, the new moon leading the way. It was only because of the trees that he remembered where the grave was. It was next to the third tree, next to the fence that separated the graveyard from the street. Across the street was a bus stop. The F stop, he believed it was, and he now chuckled at the irony.

He had never ridden a bus, but at the funeral, he had found himself looking at the women that were waiting to board it for their journey across town to work in other people's houses and raise other people's children, unaware that his Mama B was being retired to eternal rest. She too had had to raise another's child, though she hadn't had to weather the elements to do so.

There wasn't going to be anyone to remind him where she was buried or, for that matter, to bring him there, so he had focused on the details. When he'd gotten out of the limousine, he'd looked at the mound of dirt and the tent over it. Once he had spotted it, he had looked for markers. The tombstones were of no assistance, as they all looked similar and just added to the confusion. But the trees, lined neatly in a row along the fence, provided him with the perfect Polaroid. He didn't want to lose his Mama B.

Now, advancements were being made. The road had been extended and newly paved with pressed blackness. The bus stop was actually a stand, providing shelter rather than just a faded sign dictating destinations that its passengers knew all to well. He knew the buses were probably new as well but that more than likely, they still shook the ground that held her.

Though the trees were now gone because of the expansion, he could still envision them, so he found her without difficulty.

"Hey, ol' girl," said Jeremy. "I'm back. Did you miss me? I've been thinking about you. A lot. I can't say I miss you, 'cause I know you're up there watching, but sometimes I forget that I've got angels."

He looked at the ground but decided against sitting. He stood, shifting his weight from one leg to the other, like a child who had to stand in the middle of a seesaw because no one else was there to balance it. He never mentioned or spoke to this Mr. Bishop who was also under foot. He didn't know him. He looked at the plastic flowers on her grave. The sun had sapped the color from them, and he wished that he had thought to bring fresh ones.

He pressed his fingers to his lips and kissed them before rubbing them along the top of the tombstone. Again, the funeral began to parade before him. Perhaps it was the smell of his own breath that made him think of the man. He was an older man whom Jeremy hadn't seen before, but Jess and Gladys seemed to know him. The man had brought carnations and wept at the funeral like a wounded animal. At the time, Jeremy assumed he was like Miss Claire, a stranger who frequented funerals, but the man seemed to be truly grieving. He carried with him an air of tension, particularly when he came to the house after the funeral. He never came in; he just wandered outside in the driveway. Jess went out to speak to him and then he left. That was that.

Jeremy didn't understand why he thought of that man now. Perhaps it was because the man had held him near this very spot. He remembered that. Yes, many people had hugged him that day in their vague attempts to comfort him, but this man had held him almost as though he was doing it for himself. It had frightened Jeremy at the time. The cold touch of a broken man. It was his father that had said it was time to go, and the man just stared at his father. As Jeremy walked to the limousine, he looked back. The man's eyes were on them still, bloodshot and glassy.

"Who was that?" Jeremy still wondered as he walked out of the cemetery and back to the Oldsmobile.

When I was six, I had to go to several specialists. Mama B said that I was a "special child," but at the time, it was believed that I was what is now known as autistic.

This seemed to stem from the fact that I had a tendency to rock back and forth whenever I really enjoyed something or needed soothing. In such moments, no string of words could possibly be beaded together to describe my world when my swaying took control.

"There's nothin' wrong with this chile," Mama B told the guidance counselor. "If anything, it's the teacher that has the problem. He knew how to read by the time he was four, and now you wanna say there's a problem?"

I took all sorts of test with Mama B and Aunt Jess at my side.

"Which one of these animals belongs in water?" asked some man, speaking to me as though I were slow. He held up a card that had a duck, a giraffe, and a hippopotamus on it.

"The duck and the hippo," I said.

"No, Jeremy," he said. "Just the duck."

I heard something about a quack and was pulled out of the chair so quickly that my head spun. I was sent to St. Pascal's for the rest of elementary and junior high school. Mama B was right. I was special—and I wasn't the problem.

131

Several years later, it was decided that I was actually "advanced," and therefore I got to skip eighth grade, which is when I transferred to Bonaparte High. But freshman year at BH proved difficult for me in more ways than one. Besides my studies, adapting to a new environment, and fighting back, I had grown away from Mama B, and she away from me. Glaucoma and emphysema became her bedfellows, filling the void. She cut down on smoking to just half a cigarette at a time, placing the other half in the ashtray for later.

When the coughing fits occurred, she would cool herself with the Martin Luther King Jr. fan until her lungs had found some sense of freedom. She steadily flapped that fan, humming the tune of "Lord, Open Thy Unusual Door" between each cough, rocking in that rocker, next to the chifforobe where I learned about life.

She began to visit St. Francis Hospital frequently, but I was never allowed in to see her, as they had a strict age requirement and the people who knew best felt I was too young. Mama B hated hospitals because she felt that they were "playing God."

"We can watch her better here. It'll be best for everyone involved," the doctors said. But I refused to believe they knew what was "best."

I rose early the day she died. Something lifted me. Something peaceful.

"Ma Dear died last night," said Aunt Jess. She didn't mince words. Aunt Jess was always straight with me, for she knew that I appreciated it.

"What?" I asked. I'd heard her. I wasn't asking her to repeat herself. It came more as something involuntary, like a cough, a swift punch to the diaphragm. But Aunt Jess said it again, in the same way as before. She didn't say "passed" or "went over." She just repeated: "Ma Dear died last night."

"Okay," I said, and I went and climbed back under the blankets. I slept for a good long time. One soul for another.

When I did finally rise, people had already been calling to see how I had taken the news. Everyone was concerned that I would, for lack of a better description, freak out. They became even more concerned when I didn't.

"He was the closest to her," said Aunt Gladys. "I never understood it, but I guess some things just aren't meant to be comprehended."

I rarely saw Aunt Gladys. When I did, she never had a kind word for me and she had even crueler words for my father. I assumed her harshness toward me was because of him; as I felt he deserved it, I accepted it in stride.

At the wake, I sat somewhat beside myself. I was upset by the grandeur of it all. There was no doubt that Mama B deserved it, but I wasn't sure that she wanted it. She had once told me, "Don't let them go buryin' me in one of those high-costin' pieces of metal. I worked at the funeral home off and on and I know how much those things cost. Makes no sense spendin' that kinda money on somethin' that's goin' in the ground. The tribute is when you leave somethin' behind. That money should go toward the livin', not the dead."

But funerals are for the living, not the dead, so the money was spent and flowers poured in from far and wide. I walked up to look at her. I didn't feel I needed to, but I did. A gesture, I suppose, more for others than for me.

"Look at you, ol' girl," I whispered, half expecting her to sit right up out of that casket and reprimand me for being "too familiar." But she didn't. As much as I wanted her to, she didn't.

She didn't look like herself. Too made up. Aunt Gladys had seen to that. Mama B never wanted to be fussed over and she cared little about appearances. I know that.

Why did they have her glasses on?

The next day, at the funeral, it was raining. But "if you really need to go, the rain can't stop you."

The hymn was "Lord, Open Thy Unusual Door."

The sermon came from one of the oldest books, the Book of Job. I smiled at the thought.

All eyes were upon me, waiting for a breakdown that would never present itself to onlookers. Perhaps I was too young to regret, for my time with her had always been wonderful.

I sat in my newly bought suit from Selber Brothers, a suit I would never wear again, in the few family seats provided by Thompson's Funeral Home, the funeral home that used to bear our family name.

"Ya doin' just fine," said one obese woman purporting to be my aunt, though I had never seen or heard tell of her. "I know you're a big boy, but you can cry if you want."

I didn't. I didn't want to cry. Mama B's dying didn't weigh heavy on me, for I knew she had lived.

After the funeral, the cars all began pulling up at Mama B's house. To eat. The mysterious aunt, who kept trying to make me remember her, told me, while suffocating me with her bosom, that I had grown up to be a fine little man and that my suit suited me.

Aunt Jess gave me a reassuring nod for support. She knew that this part was the hardest for me.

Everyone gathered and ate, telling stale, tepid tales about the ol' girl that weren't nearly as fascinating as the stories she had shared with me. But as I sat near the chifforobe in that rocking chair, I knew I had those stories in me, and I looked down at the piece of denim on the chair and rubbed it for her.

How had Mama B taught me? I'm still unclear about that. How does the soil teach a seed to grow?

It seems that so many things take time to unfold, and all those things that seemingly don't matter to others we later find have been etched into our minds like the marks on the side of the door frame, reminding us that we have grown since the last time we checked. But if you stand in that doorway too soon, you're certain to find that you haven't grown at all, disappointment the only thing that fills the frame.

"You'll never stop learnin'," she had told me from that chair. "Black people have gone through so much that half of what we've been through never gets said, but you always keep learnin'. You learn right up to the moment you die, and then there's somethin' else to learn."

And thinking of those words, I swayed back and forth in the rocker, near the chifforobe, the place she no longer sat yet still occupied.

Jeremy heard the wailing behind him and he pulled the car over to the side of the street.

"What's the problem, officer?"

"Your headlights aren't on," said the policeman, shining his flashlight into the car. Jeremy averted his eyes from the beam. Another officer stood behind the car with one hand at his side, itching and at the ready, like a cowboy awaiting high noon.

"I'm sorry. I just left the cemetery and I guess—"

"Have you been drinking, sir?"

"Earlier, yes. I—"

"Could you please step out of the car?"

Jeremy opened the door and stepped out of the car. Between the drinks, his encounter with Jason, and the twirling blue lights, his head wasn't right. He swayed to a stumble, attempting to regain his balance.

"Can I see your license and . . . Jeremy? Boy, is that you?"

"Uh, yeah." The face was somewhat familiar. Who was it?

"Rory, remember?"

Rory? Mr. Easy Bake Oven, a policeman?

"Rory, you're a cop?" asked Jeremy, wanting to take back his disbelief as soon as he released the words. Rory was now twice Jeremy's size and his shoulders were where his neck once resided.

"Yeah. Don't look so surprised."

"No, it's just that . . . I mean, the last time I saw you . . ."

"We were just kids."

"Yeah," said Jeremy, shaking his head. This couldn't be happening. "Listen, Rory. About the—"

"Oh, forget it. I was a different person back then. I couldn't really expect someone like you to want to hang out with me. I mean, I know my grandmother always forced your grandmother to invite me over and—"

"No. I mean . . . I'm sorry about that too, of course, but—you know, I'm sorry about the headlights. You know, I had a few drinks, then I went by the cemetery to see my grandmother and . . ."

"Not another word about it. I understand. Sorry about your pop. My dad passed away last year. It's rough."

"Uh, yeah."

"He was a good guy, your dad. He did this bridge work," said Rory, opening wide to present all his teeth to Jeremy. "I guess I ate too many of those damned cakes when I was a kid. I leave that to the missus now. Got three kids. Hold on—I've got pictures."

Jeremy began to get nauseated; the blue light kept bouncing off Rory's face, making things appear to spin. Rory pulled out his wallet, then took out the Kmart Christmas photographs. Jeremy provided the idle compliments while the other cop refused to fall to at ease.

"Any kids?" asked Rory.

"No. Can't say as I do."

"I guess bein' in the big city keeps you busy. But you'd better get crackin'—wait too long and you'll—"

"Uh, Rory, listen. It's been a long day. I really—"

"Oh, sure. No problemo. It was great to see you. I'll see you at the funeral."

"Awright, we'll catch up then," said Jeremy, fully aware that his words held no ring of truth.

"Get home safe, and lock the door on the way. It ain't New York City, but we get our fair share of car jackin's. But now, down here, you can shoot 'em if they try it."

"Thanks, Rory. I'll be sure to keep that in mind."

Jeremy got back behind the wheel and the chest strap of the seat belt almost strangled him. Rory pulled his car away, waving as he passed, but the other officer's stare was piercing. When they were out of sight, Jeremy put his head on the steering wheel, trying to find whatever air he could suck in. The pictures of Rory's children, with the laser-beam lights in the background, became three dimensional in his head. He started the car, then rolled down all four windows. He switched on the headlights, and with all his concentration on the yellow strips down the center of the road, he drove on, the reflecting bumps on the strips dictating when he had strayed off center toward danger.

Without realizing it, Charles had introduced me to the world of photography as a way of making me feel less discouraged about my visit to California. Prior to that moment, I'd never been fond of photographs or photography, for they always reminded me of the image I'd never see.

I had often wondered what my mother looked like. I knew she must have been beautiful, for that's what a fantasy demands. It was Mama B who told me that there were no pictures because some people just didn't care for them. But every black family I knew seemed to dote on photographs, so her explanation made little sense to me. She said that the Indians believe that every time someone takes a picture of you, a piece of your spirit is stolen. From this I was led—or rather chose—to believe that my mother was an Indian, maybe even one of the descendants of those who founded Elsewhere.

Just that belief was all it took for me to pass up hide-and-seek and make cowboys and Indians my favorite game. But this meant recruiting others. Having been around mostly adults and believing myself one, I had little tolerance for children my age, but many children are older than they appear— in actions if not intellect.

The Baker boys lived just down the street. Their family had won a plot of land in a lawsuit. It had three houses on it, along with certain areas that resembled a car graveyard. About ten cars populated their yard; not a single one provided mobility. Some were supported on cinder blocks; other were on the ground, weeds substituting for tires. Some had the hood wide open, but others had no hood at all, their rusted wounds becoming a home for yellow jackets, lizards, and the like.

The Bakers were known to produce at the rate of rabbits. No one could say exactly how many people lived on that plot of land. I knew just the three who were somewhere near my age and the mothers who always flirted with me in that way that only older women can.

I'd walk down the street to their place and the Baker boys would be out and about. Pookie was the youngest. He was always the first to be found in a game of hide-and-seek, but he could never seem to figure out that his soiled scent was the giveaway. Chester was the middle child and mean as all get-out. He was two heads above us all and would spit so much you were surprised that there was anymore liquid available in his body. Tyrone was the eldest, so I liked him the most. But he rarely had time to hang out with us, as he was on his way to manhood.

Summers garnered notable additions to the clan, as Miss Irene's grandchildren always came from Detroit, carrying the mystique of coming from that unknown place up North. Miss Irene's granddaughter Shandra and I had been boyfriend and girlfriend ever since we had heard the words. My job was to say mean things to her, then she would do the same to me, and later we'd meet and kiss. Her little sister Precious—"to name a chile Precious is just askin' for a heap o' trouble,"—was a crier. I never knew her to do anything but cry or say, "Ooo, I'ma tellit." I believed their brother James to be the coolest thing since "sliced bread," as I had heard the older folks say. I never understood that saying, for I didn't realize that bread came any other way. But I was honored to have James as a friend, if only for the summers.

Miss Irene was probably the most popular adult in the neighborhood, for she was the huckabuck lady. A small huckabuck came in a Peanuts Dixie cup for a dime and the larger ones came in a Styrofoam cup for fifteen cents. Because Shandra was my girlfriend, I always got mine free during her visits. Or almost for free. She would steal them for me. Though Miss Irene counted her inventory closely during her grandchildren's stay, she counted only what she made, never the ones that we made while she was out to replace those we'd already consumed.

Huckabucks were either red or purple. If you timed it just perfectly, the Kool-Aid would be frozen just enough that the top was a syrupy treat that you could lap up with your tongue. I always preferred the larger ones because they required less skill to eat. The small ones often entailed the consumer's having to push them up from the bottom of the cup. Many times, the frozen treat, with a mind of its own, would shoot out of the cup to the ground. This episode was always the truest test of friendship, for a friend would understand that after you've washed it off, a dropped huckabuck was good as new.

The larger huckabucks were easier. As you worked your way down the cup, you bit the Styrofoam then spit it out to the ground. Though it hadn't snowed in Elsewhere in a "dog's year," Miss Irene's yard was the closet thing to looking as though it had, but it was a rake, not a shovel, that cleared the ground.

The notion of cowboys and Indians didn't sit well with the others. Evidently, kids in Detroit had stopped playing the game years ago in favor of cops and robbers, leaving James to say, "You all are so backwards down here."

"That shit's for punks," spat out Chester, allying himself with James. I took that statement to be some sort of Bakerism that I didn't quite understand, but I was wise enough not to acknowledge my ignorance.

Punk.

"Fine by me. We can play whatever y'all want to. Cops and robbers is perfectly fine with me. It was just a suggestion."

"Why you talk so country, like white folks? You actin' all proper," said Chester, throwing this out at me like a dart that missed the board all together but stuck to the wall, begging for a reprimand.

I had never before heard anyone comment on how I spoke. All I could muster was "That's just the way I talk." It was Pookie who saved me, for at that moment, he started to pee on himself. "At least I don't pee on myself" rolled out of my mouth with the same bitterness of a plum prematurely picked. That seemed to silence Chester.

Shandra then informed me that girls couldn't play cowboys and Indians. I didn't want to tell her that they could because my mother was an Indian, maybe like Pocahontas. I wanted to explain to Shandra, but I kept mum, as always, on the subject of my mother.

Clark soon appeared. He was their cousin and lived with Miss Irene year 'round. I once asked her why his school bus was smaller than the others; it was then that I gathered he was "special." But summers had him home just like the rest of us. Of everyone, I liked Clark best. He seemed most like me. He was bigger than all of us, and I heard it said that he was twenty. Though twenty seemed old at the time, he made twenty not seem old at all.

He did have the eyes of an old soul—like mine—and when I looked into them, I saw something that would never exist in the eyes of the others. I saw innocence, and wherever that is found, you're certain to find longing. He loved hugging, but hugs were shunned by this tribe. I let Clark hug me once. He wanted so much to embrace everyone that he was unaware of his manly strength compared to that of my eight-year-old body. He was like a child roughly stroking a cat, unable to realize that love is gentle.

Clark would often be tormented to the point of tears. It was *as though James resented Clark's presence, because it was a visible reminder that he had an imperfect relative. I don't think Clark rightly understood what was being said, but he could feel the words that were as numbing as the ice from the huckabuck, now sucked clear of color and flavor.*

It was the last day of summer—well, not really, but it was the day before Miss Irene's grandchildren were returning to Detroit, where James said they were going to have dinner that very next night with Aretha Franklin.

The day remains vivid in my mind; the cold moves on, yet the cough lingers.

It was the boys. Tyrone. Chester. Pookie. Clark. Me.

We were sitting on Miss Irene's back steps, the August heat presenting itself in a blur before us.

"Did you fuck her?" asked James.

"Fuck, yeah. Tore that shit up. Man, yo' sister know how to do it good, too," said Tyrone. "I guess Jeremy warmed it up for me, 'cuz we was all over that mattress."

Everyone exchanged soul shakes. Even I joined in, wanting to be party to the story. Palm cupped. Slide to knuckles. Patted twice with the knuckles of the other hand. Fingers then pinched and swung up to the mouth to indicate smoking reefer, then a slide over the head and back down to a thumbs-up.

James had shown us this shake on his first day. Every summer, another move or two were added. We had attempted, in his absence, to make our own additions, but they never held a candle to what James brought from Detroit.

Miss Irene's backyard was always our private place. Behind her house was a huge field that had grown wild with neglect. A barbed-wire fence separated it from her vacant yard. On numerous occasions, we had played hide-and-seek in that field, but not this summer. Hide-and-seek was replaced with the mattress in the overgrown lot, a place for sex, not child's play.

I'd been shown the mattress, but the thought of ever being on it never interested me. I had never "done it" with Shandra. But if she had not brought that fact to light with Tyrone, I certainly wasn't going to broach the matter. Maybe that was what she wanted and why she broke up with me to be with him. I knew what it was. I'd seen Aunt Jess doing it one time.

Some man was on top of her, jiggling around. She was making these uh, uh, uh sounds and so was he, but it didn't seem like they meant disdain, as is usually the case when those sounds are strung together. No, they were continuous, drawn out, the final uh always the longest, deflating to nothing.

On the day of my first sexual sighting, I had hidden in the closet in Aunt Jess's room—not with the idea of spying, for I had always hidden in certain places in the house to just get away from it all. Hide-and-seek for one. I peeked out of the crack between the closet-door panels, feeling a bit sweaty up against the plastic-covered dry-cleaned Sunday clothes that surrounded me.

When she entered the room with the man, I dared not present myself. I was trapped there. They undressed, and he climbed on top of her but didn't cover himself with the blankets. I didn't quite understand why she would want this man on top of her like that. The look on her face was as though what he was doing was painful, but he kept right at it. I wanted to jump out and protect her, but when he asked if she liked it, she said that she did. When she opened her eyes, I had to wonder.

With that long last uh, he rolled off of her onto his back. She stroked his chest a few times, but just a few, because he popped up from the bed. "I gotta get goin'," he said, motioning for his boxers on the floor.

"You wanna wash up?"

"Nah, I'll do it later. I gotta get goin'. I'll call ya later."

Aunt Jess stayed in bed and watched through the window as he went out the front door. I sat back in the closet, not *wanting to see any more. I thought she would stay in bed forever. It wasn't until I heard the bathwater running that I crept out of the place I'd intended for solace, soaked with sweat, feeling I should wash up, but for a different reason.*

Still, I couldn't imagine Tyrone and Shandra doing that. The thought of it made me dislike her even more and hate him for his lack of discretion. It seemed peculiar to me that James didn't mind that his sister had done this. It was as though he was proud, not of her, but of Tyrone. "Hey, if you can get them drawers, get 'em," he said to Tyrone. They all laughed, even Clark, but I couldn't imagine that he had ever done it. I doubt he even knew why this was funny.

"Yeah, Jeremy, you missin' out now, man. But I 'preciate you handin' her over to me," said Tyrone, hand cupped over his crotch. I couldn't help but wonder if indeed I had really missed something. "Bitch wouldn't suck my dick, though."

"Ah, man, she wouldn't suck yo' dick?" said James, her own brother. "Nothin' beat gettin' yo' dick sucked."

They all concurred, as though sex and its many acts were as common to them as biting off the Styrofoam that kept the huckabuck from making cherrylike stains on cotton garments.

I too nodded my head, yet not in agreement. I put my hands in my pockets, pushing down as far as they would go, grabbing for anything to distract me, but all I found was lint.

"Don't go puttin' yo' hands in yo' pockets. It's too late to be gettin' hard now. You could still be pokin' that, but you were too busy ackin' like a l'il ol' punk, so I had to step up to the plate. Batter up!" he said, laughing before adding, "Hank Aaron ain't got shit on me."

Punk.

Tyrone had a responsive crowd now, and they followed his lead, hooing and hawing, even Clark. I took my hands out of my pockets posthaste and though I wanted to say something in rebuttal, like "Fuck you" or "Yo' mamma" or one of the many puerile insults that I'd heard thrown around, I knew that coming out of my mouth, they wouldn't carry the same vituperative heft.

"You wanna see somethin'?" asked James. Our excitement was piqued for a moment, for the way he posed it was as though he was going to show us the most amazing feat in the world. If he was going to be supping with the queen of soul, then anything was possible—and it had to be if it was going to top Tyrone's grandstanding.

The yeahs of everyone followed—everyone except Clark, for he rarely answered. The only sound I'd ever heard him make that wasn't laughter was like a yelp that came from deep down. I thought of it as a baby crying in the night, frustrated because no one could understand what the problem was.

"Come on." James shot off the step. He went down to where the middle line of the barbed wire had been cut out of the fence. Tyrone placed his foot on the bottom line, then pulled up the top one as we all stepped through, free of puncture. We walked through the field, me looking down rather than ahead, certain that a ground rattler or some other creature of the grass was ready to defend its home. Later on, I knew I would have preferred that.

We got to the mattress. I wondered how many others had been on it, whose stains covered the striped fabric, and did they too just roll over and wash up later.

James looked around to see if anyone else was in view, as though what he was about to show us was going to be mind shattering. When he saw that the site was secure, his hand found its way to his zipper and he slid it down. With the thumb of the other hand, he pushed down his underwear and kept the zipper's metal teeth pried open. He took his other hand off the zipper and pulled out his dick. There it was, without a thought, privacy relinquished. He began to wiggle it up and down, and in no time, it began to inflate like an inner tube in a bicycle tire until it looked as though it would pop.

"*Watch this,*" *said James, as he sat Clark on the mattress. He stepped up onto the mattress, a hand on Clark's shoulder so he could maintain his balance on the worn springs. Clark just sat there, watching us all. James put his dick near Clark's face, and as though they had done it hundreds of times before, Clark took James into his mouth, just like a child sucking on his thumb.*

None of us made a sound. I had unknowingly stopped breathing. Clark's eyes just wandered from corner to corner, not at all out of embarrassment or to see if anyone was approaching. They just wandered because that's what they always did. When he looked at me with those old eyes, I wanted to see a tear fall, but none was to be seen. I wanted to do something to stop James, but I couldn't move, couldn't speak. The air was thick and lost, as if a belt squeezed to the last hole was around my neck, making my heart pound frantically. James put his hands on the sides of Clark's head and began moving back and forth. I was not to escape to the comfort of a rocking chair.

"*James!*" *It was Miss Irene. All heads turned at lightning speed. We expected her to be standing over us, but she was calling from the front of the house. James looked up and tried to pull himself away from Clark, but Clark wouldn't let go so easily. James punched him on the side of the face and finally, like a baby burping, Clark released his suction. He looked up at James. The punch was more devastating to him than the deed I had just witnessed; yes, at last, the tears came.*

"*James!*" *screamed Miss Irene again.*

"*I'm comin'.*"

"*Hurry up, then. Ya momma's on the phone.*"

James zipped his pants and began running back through the field, with the Baker boys closely in tow. Clark tried to find some spring from the mattress to get up, but he was so big that leverage weighed against him.

There we were, together. I walked over to him, trying to help

*him up, and he reached for my zipper. "No! Clark—no! Don't
do that!" My scream shook every strand of overgrown vegeta-
tion in the field, and Clark again began to cry. I sat down
on the mattress, forgetting its filth. I stroked his back, telling
him it was all right, the way Mama B or Aunt Jess had done
to me every time I cried. He began to rock back and forth as
I always did, and after a while, he stopped crying. I stood,
and it took all my strength to help him from the mattress. We
began to walk through that field. He grabbed my hand and
I let him hold it, and I didn't care about snakes or anything
else. We just kept walking.*

*True to form, I never considered speaking of what I had
seen. Though they were all too old to play cowboys and Indians, they still used the word* tattletale *as if it was Mr. Webster's
finest. I didn't want to be likened to Precious. I didn't know
what the word* punk *meant, but I knew that if it meant that
I wasn't like James, then I didn't at all mind being called
that word.*

*The next day, Miss Irene sent her grandchildren on their
way. Because my life had been filled with things not said and
frequent good-byes, I had become accustomed to it. But never
had I been so happy to see someone leave as I was when
James left for Detroit, where they played cops and robbers, not
cowboys and Indians.*

*A few weeks later, I came in from school and Mama B was
in the kitchen, but she wasn't making cornbread for me. She
was frying fish. I always considered fish to be our Friday meal,
but it wasn't Friday.*

*"We're having fish today?" I asked, brushing against her
side. But she remained distant. No hug hello or questions
about my day.*

*"No, Patience, this ain't for us. I'm makin' it for Miss
Irene."*

"Is she sick?"

"No, she got a bit of bad news today. I'm just takin' somethin' over to ease the burden."

I knew Mama B wasn't telling me the whole story. She always tried to protect, as was her nature. The truth always came from Aunt Jess. I knew the only two reasons someone took someone else food was when they were ill or someone had died. Since Miss Irene wasn't sick . . .

I ventured out of the house in search of answers, not at all caring that I would miss Flipper's adventures that day. I went to the rental house to talk to Miss Claire; if death was involved, I knew she would be privy to the information.

As it unfolded, James had been associated with a gang back in Detroit. "He was killed dead, just like his no-'count daddy," said Miss Claire. "I don't know how his mamma kept him outta trouble this long. Now, she could send them down here to Irene, Lawd bless her soul, for the summers, tryin' to get him away from that foolishness, but who was watchin' him the other nine months? I truly feel for Irene."

It was different for me to think that someone thirteen years old, the age of manhood in certain cultures, two years older than me, could be shot. I supposed Detroit was different and we were backward down here. But if this was backward, it was what I wanted.

I can't say I was disappointed by the news about James, for though I couldn't envision him on the ground with bullets in his body and bloodstained clothes, I could remember Clark's eyes.

When Mama B returned from delivering the food, she was out of sorts. She sat in her rocker as if even its frame couldn't support her, an oak turned into a willow.

"Are you all right, Mama B?"

"Yes, Patience."

I could tell by her posture that she didn't want to speak any more about it, so I let it and her rest. I went and sat on the *back steps and I looked out into the day. Later that evening, I found Mama B in better spirits.*

"Mama B?"

"Yes."

"What's a punk?"

She stopped her brush midway through the length of her hair. She brought it down to her lap and pulled off the excess from its teeth. She balled the hair between her fingers then placed it in the ashtray and struck a big wooden match, setting the oiled hair ablaze. A sizzle and the smell of burn filled the air for an instant. When the hair had shriveled before me to nothing, she blew out the match. Holding it up she said, "This is a punk, Patience. It's a piece of wood—just a piece of wood."

chapter 11

❦ ❦

Right before pulling the car into the driveway, Jeremy turned off the headlights so as not to stir Jess or draw too much attention to himself. It was after midnight and the day had been longer than the time within it. Tension crowned his shoulders like thorns. He rolled his head on its axis, trying to loosen the kinks of days and years that went away for short periods but always returned, embedding themselves in the muscles and demanding attention.

"Hey," said a voice as he got out of the car. It was the rental tenant. She was sitting on her front steps, having a cigarette. Jeremy walked across the lawn to talk to her. Each step made her appear more alluring—a stranger in a familiar environment.

"Oh, hello. I didn't see you sitting there. I'm Jeremy."

"I know who you are. Ever'body in town knows who you are."

"Well, I wouldn't go that far."

"You're a lot better lookin' than I expected."

"Excuse me?"

"You know. I thought you were some nerd or somethin'."

"Sorry to disappoint you."

"If I was disappointed, I wouldn't've called you over."

She took a long drag on her cigarette, never moving her eyes from Jeremy, once again giving him a once-over, without a hint of subtlety. Jeremy looked away from her eyes, focusing his attention on the little house where he'd lost his virginity. Now here was this young woman, sitting on the very step he had crossed to venture inside for that momentous event.

"Can I bum a butt?"

"What?" she asked coyly.

"A cigarette."

She handed him a cigarette, then passed hers to him to use as a lighter.

"Um. Menthol? Thanks. I quit, but . . ."

"Yeah."

"So, what's your name?"

"Kim."

"Well, it's a pleasure to meet you, Kim. Kids asleep?"

"How you know I got kids?"

"Uh, well, Aunt Jess mentioned it, and I saw them out here earlier today," said Jeremy, trying to get himself out of the way of his tongue.

"I guess she told you I was behind in the rent, too."

"No. She just said that you were the new tenant and you lived here with your kids."

"Did you ask her about me or did she just tell you that?" asked Kim, taking another drag on her cigarette. Her dress dangled between her legs and her toes, bare, curved around the lip of the step. It had been a long day. Jeremy was tense, he couldn't relax.

"I asked her," he said. "I saw you when I pulled up this afternoon."

"Yeah, I bet. And what were you thinkin' when you saw me?"

Jeremy sat down on the step next to her, trying to enjoy the menthol cigarette, but it had become nothing more than a prop, like in a photo shoot.

"Oh, I don't know," he said. "I was just curious, I guess. New face, you know. I was very close to one of the tenants. He was like a father to me."

"Well, ain't no fathers been up in here lately, that's for damn sure."

They both sat there on the step, taking in her statement in

their own way. Jeremy had longed for the day when he could be a father. For a few seconds, he could envision his being the father figure for the kids he'd yet to meet, in that house. Just knowing how much it would mean to them to depend on someone made him feel fond of the young woman, nine years his junior. She opened and closed her legs, grazing his every so often. As she did so, the fabric of her dress straightened and concaved, rising well beyond her knees.

"You wanna come in for a minute?" asked Kim, throwing her cigarette out in the yard. Jeremy looked over to the house where he knew Jess was and then back at Kim. He rose from the step without a word and stepped on the cigarette that she had just flicked away, and then picked it up. With his fingers, he pushed out the lit tobacco of his own until the glow, like a meteor in the night, fell to the ground. He stepped on his as well and then returned to the step. But he didn't sit. He stood in front of her. With that, she stood and then opened the screen door, and he followed her in.

Charles was the first man I can say I ever truly loved. Not loved because I was supposed to, but loved because he warranted it.

It was in his house, at fourteen, that I lost my virginity. Rolanda Watts, the original get-it girl, had moved into the neighborhood. Her reputation for being an easy lay preceded and followed her, and I felt privileged to have her in near accord.

We sat on the steps of Charles's house one sultry day when the Southern heat reminded us that we were mere animals and scents dominated. Charles had given me my own key to his place so when I needed a refuge I could come over and play records or "what have you,"—although I don't think he truly knew what "what have you" would entail.

Rolanda and I went in and sat on the sofa. She wasn't the sort that you had to coax with an offer of a soft drink or music. She was strictly business. When she asked if I was a virgin, I denied it, of course. No boy would admit to such a thing, fearing the information could be used against him. Once she was convinced that I wasn't lying, the necking began, followed by the petting. She unbuckled my belt, then unsnapped the waistband of my pants. Her hand found its way inside my pants and under the elastic of my underwear. I felt her gripping me. I felt her tongue darting in and out of my mouth, and I tried to catch it with my own. She lay back on the sofa, bringing me on top of her. With her foot, she pushed my pants and underwear down to my knees, and her pants and underwear followed suit.

I supposed that sexual organs were smarter than they are. For some reason, I just thought they would pop into place like a TV stand. I kissed the left side of her neck, working my way up to taste the inside of her ear, then down to the closed lid of her left eye, then the right eye, then the right ear, and back down to her neck. How I came to choose this path, I'm not sure, but it just seemed to happen without direction. Her hand again found its way down to where the heat rose, and she slipped me into her moist insides.

My pants and underwear were still at my knees, cutting off my circulation and restricting my movement, yet I could feel myself, within, without. I couldn't explain the sensation at the time, for I was awestruck.

I moved myself in and out and her hands turned my T-shirt into an accordion as she moved them up and down my back. I obliged with the appropriate notes. I could feel her fingernails indenting my flesh. Though it was somewhat painful, the prickling didn't deter me. I could feel her moist pubic hairs dancing with mine. I didn't speak—the thoughts in my brain were too fragmented to be properly assembled, their language foreign.

I didn't look at her, fearing this was just a notion and she wasn't really there, but I could feel her under me and hear *her, the noise of the sweat of our joined stomachs adding to our gibberish. I remained focused on a spot on the wall. Even when I did look down to make certain that I was indeed doing this, it was only for a second of reassurance, then my eyes flew back to the spot on the wall as though instructions had been scrawled there to keep me focused on our rhythms.*

Finally, all my weight found itself on her. My bird's chest pressed against her well-formed breasts and my head rested on her shoulder. I stroked her hair and kissed her neck, wanting to establish affection, but her hands were no longer caressing my back. It was as though her body was absent. "Get off me. I'm hot," she said. I did as she asked, and she sat up and pulled up her panties and pants. I followed her lead, finally pulling mine up as well. "I gotta go," she said. "I'll talk to you later."

That said, she stood up. I tried to kiss her, but she put her hand to my chest, halting me. I tried to grab her hand to lead her to the door. The moment I did, she shook it loose. No more words were spoken. Just as she left the front yard, Charles pulled up. He got out of the car. He didn't say anything, but he watched her brazen walk down the street.

"Hey, Chuck," I said, trying to provoke a sense of man to man.

"I'm not amused," he said, carrying in his grocery bags from the car. "Here. Make yourself useful."

I grabbed the bag and we went into the house. He scanned the sofa without stopping. I too looked at it to see if any hurly-burly clues had been left in the gulches, like fallen coins. Oddly enough, I wanted him to see clues. I wanted to brag about it. I wanted someone to know aside from Rolanda. Charles refused to accommodate me. He just left me there with my downsized thoughts.

"I hope you don't mind that I had someone over," I said, trying again to initiate conversation, hoping to lead it in the direction I wanted it to go. He said nothing. "She just came

*by to hang out. You know—listen to records, stuff like that."
Charles looked at me and smiled, his face wearing his best
I'm-not-interested look.*

*When his groceries were put away and the brown paper
bags were neatly folded and tucked between the wall and the
refrigerator, he said, "Come here for a moment." I followed
him through the house and into the bathroom. He opened the
medicine cabinet. "Your Aunt Jess thought that I should speak
to you. I see sooner would have been better than later. I bought
these for you," he said, but the instructions that should come
with them were in his eyes, for the only words that followed
were "These are always here. You've got too much to offer and
too much to lose. Remember that. Being a man is one thing,
but being a father is something else. If anyone should know
that, it should be you."*

*He handed me the box, still in its shrink wrap, and walked
out of the bathroom. I stood there looking at the package, but
found no irony in the word Trojan. I put the box back into
the medicine cabinet and closed it. I stood there looking in
the mirror. I looked the same—to my surprise, the pimples
had not magically disappeared.*

*Charles put on the cut "Tryin' Times" from the First Take
album. He loved Roberta Flack. I'd heard him tell some of his
friends, "The '69, '70, and '71 albums, now, that's when the girl
was showin' out. But when you're feelin' downright angry and
you think you wanna spit on somebody, there's nobody quite like
Nina." Roberta and Nina Simone were always on the turntable.
"Cleanin'-house music," he called them.*

*He fixed himself a drink without offering me anything,
which was unusual. He walked out of the kitchen. Right as
he was about to sit on the sofa, he stopped himself and moved
over to the loveseat.*

*"So. How was your day?" he asked with a catlike grin, and
all was back to normal.*

❖   ❖   ❖

The three kids were asleep on the couch, bodies contorting in a way that only children's and cats' can. They looked pure and peaceful and were oblivious to Dionne Warwick's invitation to call in for a psychic consultation.

"I don't get any peace until I get them down," said Kim.

"I can imagine they're a handful."

"It's not too bad. I love these boys. They're my life."

The house was a bit disheveled; it wasn't as Jeremy had remembered it. He wanted the memory of Charles to be preserved, in more ways than in his mind. Kim walked over to him and put her arms through his and around his waist. She rested her head on his chest, then lightly kissed him there, working her way up to his lips, once, twice, then his mouth opened to oblige. Her body felt warm next to his and their kiss intensified as his hands explored her body, finding their way to her ass.

"Momma?" said one of the little boys, standing up from the couch.

"Gwon in the room and go to bed," said Kim. She never took her arms from Jeremy's waist; she just held him. The little boy walked toward the bedroom, but his eyes remained on Jeremy. He stood at the door and flapped his fingers in a wave. Something in that motion was sobering.

"Uh, I really should be going," said Jeremy, nearly running away from Kim.

"Why the hurry?" she asked, trying to reassure him. "It's okay. He's fine."

"I know. It's just . . . you know. It's been a long day. The flight and all."

"Oh, I see how it is. I'm not good enough for you."

"No, it's not—"

"Whateva," she said, flopping down on the couch. "Tell Jess I'll try and pay the rent when I can."

Jeremy pulled out his wallet and held out some bills. She looked at him for a moment, yet her face held no question. He felt the weight of his outstretched hand and it began to falter, as though he were holding a tree rather than a product of its timber.

"Thanks for the cigarette," he said.

With that, she took the money, folding it quickly in her hand without looking at it. "You're welcome," she said.

I was in the yard when the two cars pulled in front of the rental house. The policemen stepped out, looking at both houses, then at me. Two of them came over to our house. I could see the front doors flick open all over the neighborhood like shutters sending Morse code messages. Porches became mezzanines for the show that was our yard. Mama B and Aunt Jess came outside.

"Hello, Mrs. Bishop. How you doin'?"

"Fine, Tom. Is there a problem?" she asked, looking at me as though I might have strayed like many boys my age do.

They pulled Mama B and Aunt Jess off to the side, nearer to Charles's house. I wanted to walk over too, but I knew my place in moments like that.

The officer presented a plastic bag and I could see a wallet inside it. From his pocket, he revealed a photograph. I saw Aunt Jess grab the photo, the paper quivering in her hand, but no breeze was to be found. I saw the policeman's mouth move, and Aunt Jess nodded her head. The cop said some words and the photo fell to the ground, like a feather in the wind. The heat of the day rained down on them, proving that the sun too could bring lightning, striking at will.

The nearest thing I ever had to a father had been taken away from me. The man in my life was decidedly not man enough to "someone or someones," as the policeman had so politely phrased it.

Charles and another man were found by some teenagers in the woods of Crossing Pointe. CP, as it was known, was a common area familiar to everyone—or at least to most—for the many firsts that happened there. It was where the moon shone bright through moss-covered trees and across the marsh of the lake, while nonjudgmental egrets watched. It was so untamed that the landscape seemed to provide protection from the rest of the world. Uncounted campfires had been lit and stories told over them. Virgins had left otherwise, joints were smoked, and beer and Boone's Farm were swilled. Many good times were had at Crossing Pointe, but all of that was now shadowed.

Though the details never made it to the papers, word got out that their maleness—not their manhood—had been cut off and placed in their mouths. A single bullet to each of their heads.

That was my last vision of him.

A man who had loved me, unconditionally. He was, in a sense, my father. Man enough to be that, yet not enough for others who couldn't possibly see beyond the trees.

"Umph. Serves him right," said Mrs. Richards, standing in our yard later that evening. She came empty-handed, refusing to bring any nourishment to assuage our pain. All that accompanied her were words to gnaw at the flesh. "Sister Bishop, I know you couldn't've known that sort of devilment was going on in that house, but we all knew—and knew it was just a matter of time before the Maker had had enough."

"Well, the Maker has let you live a long life, so your theory doesn't pan out," said Aunt Jess with a serpent's tongue.

"Would be like you, Jessica, to defend him. People been wonderin' 'bout you, too. Ain't right for a woman of your years to not have no children."

"I don't have any children 'cause I don't have a husband. But that didn't seem to stop ya daughter, did it? The Maker didn't say anythin' about that, did He?"

"Jessica," said Mama B. Her tone was soothing, but a reprimand nonetheless. "Patience, gwon in the house, please."

I stood there staring at the woman who called me ugly boy.

As the saying goes, I wouldn't have spit on her if she were on fire.

"It's just downright shameful that you had him right across the way from your grandson. I used to see them playin' together out in the yard, him always touchin' the boy. Oh, yes, I was watchin'. They just laughin' and carryin' on, not to mention all those wild friends he had goin' in and out of there."

"Patience, in the house."

I did go in the house.

"Close the door if you would, please."

I closed the door.

I watched from the window. Though Mama B had been ill, life came into her, and not once did she reach for the bottled breaths.

Mama B and Aunt Jess became blurs and the swiftness of their movements seemed to blow Mrs. Richards back across the street. I don't have a clue as to what was said, but I knew that the Maker had nothing to do with it.

They both came back into the house. I sat with wide eyes and anxious ears, hoping to hear what had transpired.

"The woman makes you want to lose your religion. It's not that I wish bad on her; I just want her to die, that's all."

"Jessica, that's enough," said Mama B. "That's enough."

A man came to the door. He was introduced as Charles's father. He'd driven in from Vicksburg to collect his son. Charles had once mentioned that he was from Mississippi, but that was all I knew. Like so many who move away, home wasn't a big topic for him.

The rental house had remained the same. Only the police had been in it; the tape they rolled around it made me think of a tree wrapped in toilet paper during Halloween. When Charles's father arrived, Aunt Jess wasn't home and Mama B

told me to go over and let him in. "I just can't go over there,"
she said. "It's too much for me."

His father walked into the house and looked around as
though he expected to see Charles.

"So this is where he lived?"

"Yes, sir," I said.

He walked around for a bit more, his hands touching different surfaces as he did. He sat on the sofa. He put his head
in his hands. Ashen and worn, they covered his face, but I
could see the anguish peeping through the appendages.

"No matter what anyone says, I . . . well, uh, he was like
a father to me. You . . . you should be proud of him," was
all I could think to say. The words spun through the house
that was once brimming with life, and when they had made
the run of the place, they came back and stood with me, a
successful boomerang pitch. His face was still covered, so I
didn't notice a single tear until I looked at the floor.

"Something's in my eye," he said abruptly, springing from
the sofa. He walked over to the bookshelf and pulled down
Charles's Bible. He held it in his arms. I had seen Charles
read that Bible many times. He told me it was the last thing
his mother had given him before they parted ways. He kept
her photograph in it as a marker. It was Charles who had
convinced me that there was a God.

This occurred after my confirmation. I had been taking the
CCD classes before Mass to prepare and was truly concerned
about any form of confession. I asked my teacher, Mrs.
Fortenberry, if Father would be able to see us when we made
our confessions. She assured me that he wouldn't, that it was
strictly confidential and my sins would be between only me
and God. This threw me off, because if only God was going
to know my sins, I wondered why I had to deal with this
middleman.

Finally, the day arrived. Mama B, Aunt Jess, and Charles
were beaming at me all decked out in my little white ensemble.

Mass started, all the kids went through the ceremonial proce-dures, and then it was time for confession. To my surprise, for our first confession we weren't going to use the confessionals like the adults. Instead, we were going to use a room off to the side. I was hugely disappointed.

When my turn came, I walked into the little room and stood in front of an altar with a screen attached to it. I couldn't see Father, but I assumed he was behind it. The screen was about two feet wide, just large enough for one person to fit on each side. As I had rehearsed, I kneeled and made the sign of the cross.

"In the name of the Father, the Son, and the Holy Ghost. Forgive me, Father, for I have sinned."

He barked out his dogma, and as I was just about to start telling him things that I thought could possibly be considered sins, I glanced to the right. There on the wall was a full-length mirror, hung just so that it focused the screen. With the slightest glance, he could see me and I could see him.

Mrs. Fortenberry had lied to me. The Church had lied to me. And I felt that if they had lied to me, then this God must have lied to me, too.

I can't remember what I said during the confession. I mum-bled something just to get it over with. I stood up and left the room. None of the other kids had seemed to notice the decep-tion, yet I had. I had to be the one to notice, the one to figure it out. Wasn't I the smart one? In that moment, I had lost all faith, remembering that even the pope rode in a bullet-proof limousine.

Still clothed in melancholy, I later went over to visit Charles to relay the story. He was appropriately reading his Bible. I told him I didn't believe in God anymore. I had already seen and experienced too many things that drew me back to that question that I had asked Mama B: "Was God sleeping?"

Charles closed his Bible, but he used his finger, rather than the picture of his mother, to mark his place. He rubbed the photo, and I almost hated him for having it.

"Always believe, J. I believe in God because I believe in the goodness that exists within the unknown. All the Christians in the world can say what they want, what they will, but they can't say they truly know God. God is nothing more than unknown goodness. It is man or sometimes woman that makes Him anything less than that."

He put the picture back in the Book, then closed it and placed it back on the shelf. That was all he ever said about that.

Now, his father stood there holding that very Bible, and the tears did flow freely. This time, he provided no explanation. All he said was "No matter what, that was my boy. What in God's name have they done to my boy?"

He took the Bible—"His mother should have this now"—and a few other things. He told me to donate the rest to the Salvation Army.

I alone packed up the house. I had helped bring those boxes in; now I was taking them out. Everything was donated, except for the camera. I kept that. It was the first of what would be many cameras, yet it always seemed appropriate that the first would come from Charles.

When I locked up the house, the curtains were still in the windows. Mama B had me go back over and take them down. After that, the house truly looked barren. Later that evening, I saw her cutting a swatch from one of the curtains, another patch added.

chapter 12

∾ ∾

"**I** thought you were gonna sleep all day," said Jess as Jeremy dragged into the front room. "There's coffee if you want some."

"You are a savior."

"You just now figurin' that out? How was Paul?"

"Same as he ever was."

"He's a good kid. You know he's started up a program for foster kids."

"Yeah, he told me. Beth said he loves it."

"Well, if that girl over there don't watch it," she said, gesturing toward the rental house, "her kids are gonna end up there. Fast as she wanna be, that one. Steal you drawers with your pants still on. But you know how I am, so I try to keep an eye on her and watch out for the kids when I can, but it ain't easy."

"I saw her last night on my way in. She told me to tell you she had the rent."

Jess looked at Jeremy but didn't question him. She let him off the hook by changing the subject. "Heard you saw Carol last night, too."

"My goodness. You can't do anything around here without someone knowing," said Jeremy, certain that Jess had seen him going into Kim's house.

"What are you gettin' all riled for? She called this mornin' to tell me about the funeral arrangements and she mentioned that you stopped by. Ain't no sin in that, is there?"

"Sorry, I didn't sleep well. Can I bum a cigarette?"

Jess pulled out her cigarette case and the lighter that was tucked in its side. She lit a cigarette and passed it to him.

"I'm tryin' to quit," she said.

"Me, too."

"I usually smoke half of one, then put it out and smoke the rest later."

Jeremy inhaled deeply, pleased that it wasn't menthol.

"I saw Rory last night. He pulled me over for not having my headlights on."

"Funny. He told me you had been drinkin'." Jeremy again looked disgruntled for a moment, then he let go a smile. "He just wanted to make sure you made it home in one piece. Brought a cake over."

"I wasn't drunk."

"Nobody said you were."

"I just had a couple of drinks at Paul's. No big deal."

"You don't have to explain anything to me. You're grown. But I do think we've lost enough people around here, and I don't think I'm ready to lose no more just yet."

"Okay."

"You went to the graveyard?" she said, watching him closely. "You know you were her life."

"I just thought I'd say hello."

"She loved herself some you."

"I know."

After his third cup of coffee and twice as many cigarettes, the night before surfaced through a hangover. He thought of his confrontation with his brother.

"Do you see Jason much?" asked Jeremy.

"No. He's too busy comin' into his own. They'd been havin' some trouble with him. Arrested once."

"For what?"

"Started some fight over nothin'. The chief told ya daddy to let him stay in jail overnight, that that would cure him. But ya daddy wasn't hearin' that. Said no son of his was gonna stay in jail overnight if he could get him out."

"Superdad to the rescue."

Jess chose to ignore this statement but said, "You know, ya daddy spent the night in jail when he was Jason's age. He never forgave Papa for that. He knew that Papa could get him out if he wanted to. We knew everybody at the station."

"What'd he do?"

"Turned out, nothin'. They said he shoplifted somethin' from Scalia's. The store was busy and he just wanted a candy bar, so he put the money down on the counter and walked out. Scalia's son, who was just a bit older than ya daddy at the time, was runnin' the store 'cause his parents were on vacation. Well, he called the cops and said ya daddy had stolen the candy.

"They came and picked him up, and though ya daddy denied it, Papa didn't believe him and said he had to pay for his actions. Of course, Ma Dear pleaded with him, but he let them take him down anyway. It all came out in the wash later. When the Scalias got back, they heard about it. They had known ya daddy since the time they opened the store—and they also knew their own son and his ways.

"He later told them that 'maybe' ya daddy had put the money down on the counter. Mr. and Mrs. Scalia brought their son over and came right to that door there. Papa came out of the rental house when he saw them drive up. They made their son apologize to ya daddy and offered to have him mow the lawn for a month to make up for it.

"Papa told them that it was awright, that 'boys will be boys.' And that if he hadn't been in such a hurry, this wouldn't have happened. 'He should have waited and gotten a bag, and that would've saved us all some trouble.'

"Ya daddy never forgave him. They rarely spoke a word to one another after that, tryin' to avoid each other. Not that they ever had a good word to say before then."

Jeremy was glued to his seat. That a coffee cup was in his hand had escaped him, so when he did come out of the daze that the story put him in, he raised the cup to his lips and found its contents cold.

"Jason is just lashin' out. That's all. He's a good kid, deep down. So was ya father. So were you. There's a lot about ya father you don't know."

"I'd say it's a bit late for him to share those stories with me now."

"For bein' so old, you still actin' like a child. Ya father ain't even in the ground yet and you still can't find a decent word for him," said Jess, getting up from the rocker and going toward the kitchen.

"So his being dead is supposed to change things? Is that what it is? I'm supposed to just forget everything?"

Jess swung back around with all her years behind her.

"No. But if all you can remember is the bad, then maybe you had a part to do with it. You think you're the only one around here that suffers or was hurt? Why don't you stop feelin' so damn sorry for yaself and think of others? Believe it or not, the sun does shine before Jeremy Bishop gets up."

"Oh, I see. We're supposed to put on our blinders and forget that he dropped out of my life, then ten years later decided he should be a father and include me in his new family?"

"At least he made an attempt to come back. That's more than most ever do. Most of the time, once they gone, they gone. And you ain't never been out of his family. You were with us. We're his family. Maybe you've gotten up there with your fancy life and just forgotten about everythin' else, like we didn't exist, but how many times did you call him? How many, J? Did you ever make one attempt to reach out to him? Every try he made to reach his hand out to you, every single time, you closed the door on him and tried to fault him for it when you did."

"What was I supposed to do . . . ?"

"Grow up! That's what you were supposed to do. I wanted you to grow up to be the man we raised you to be. That's what I wanted you to do. You think I wanted to sit in this house forever, raisin' my brother and sister and takin' care of Ma Dear? No. I was young once. I wanted to see things and places, but I had responsibilities, and in time, those meant the world to me and I accepted them without a second guess. I accepted that, J. And not once did I regret it. Just go somewhere, anywhere. I don't even want to look at your face."

Jess went into the kitchen. The pots began to slam and a dish shattered. Jeremy couldn't move. He had been stung, and his entire body was numbed by the shame coursing through his veins.

On my fifteenth birthday I received a car.

I went through my freshman year at BH without one, bumming rides to Taco Bandito, McDonald's, and other high-school lunch haunts in Elsewhere. But when you're fifteen, the gift of a car seems as common as puberty.

Some kids found themselves with new cars, but most families handed down the Volvos, the Cabriolets, the Jeep Wagoneers, the BMWs. Mama B and Aunt Jess had been hinting that I was to receive the 1962 Rambler that had been parked in the driveway, next to the Lincoln Town Car that had replaced it. Mama B never learned to drive; she was always driven. The Rambler was a classic car, and I had dreamt of the day I would graduate from the passenger to the driver's seat. She had manual everything. I called her Betsy—not for any particular reason, aside from the fact that I was Southern and Betsy was a common name for an old car or a mule.

Betsy was a blue four-door with a white top. With akimbo pride and vested interest, I would scrub the top, ridding it of the black spots that were evidence of the pecan tree's fertility, because I knew one day she would be mine.

My fondest childhood memories are of whenever Aunt Jess

*started up that car. It took me away from my world. But I
was bereft: right before my birthday, the Rambler was sold. I
took this to mean that I was going to get a new car. New cars
weren't nearly as acceptable for boys as vintage ones, but if it
was going to be a Mazda RX-7, I was ready to accept it. I
had hinted to that very effect, and Aunt Jess and I had driven
by the Mazda dealership.*

*The morning of my birthday, a carrier truck pulled up in
front of the house. The driver came to the door, wearing a
smile. He asked if I was Jeremy Bishop. I told him that I was.
He said that he had a delivery for me. I went outside and
stood in the driveway. Aunt Jess was at my side, and I could
see Mama B looking out the window next to her bed. I knew
it was a car. What else could it have been in this huge
eighteen-wheeler?*

*The man opened the cargo door, and though I wanted to
sprint over for a look inside, I remained patient. He pulled down
the ramp and I heard the engine start. As he backed out, I
noticed that the car was small and blue, just like the Mazda I'd
picked out. I was ready to christen it Betsy II. But as it inched
its way out for full inspection, I saw that it was the toy car
I'd seen so many years ago. It had a big ribbon on it.*

*"Well, it's your lucky day. I know a lotta kids would die to
have a car like this," said the driver. "Just sign here and she's
all yours."*

He was right. When I saw it, I almost died.

*Aunt Jess told me that it meant a lot that my father had
given it to me. But my pride refused to allow me to hear her
praise for him.*

*I drove the car for a few days. Everyone seemed impressed
that my father had sent it to me all the way from California.
Everyone but me. The compliments never made it better. One
day, I returned without it and pulled up in another classic,
the one I'd wanted originally. The Rambler. My Betsy.*

Jeremy was in the yard, swaying on the swinging lawn chair, when Jason's car pulled up. His argument with Aunt Jess had exhausted him, but he braced himself for another round, need be.

"Wassup?"

"A preposition," said Jeremy. He couldn't resist his own tart tongue.

"I-ight. Good one. I'll have to pinch that one from you."

"Where'd you get that word from?"

"What word?"

"*Pinch.*"

"I spent a month in London last summer with the study-abroad program at Riverwood. We'd go in and pinch stuff from the stores 'cuz ever'thang was so damn expensive. Steal, you know?"

"Yeah, I know what *pinch* means. I just was surprised to hear you using it. I guess I should just call you Gulliver."

"Well, the favored son isn't the only one that gets around, you know?"

"We're not gonna start that up again, are we?" asked Jeremy, releasing a big breath. "I'm really not up for it right now."

"Nah, bro. It's cool. Actually, I came over to apologize," said Jason, pulling out his Newports.

"Can I—"

"For someone that's so successful, you a bummin' muthafucka. Here. I thought you might need your own." Jason handed him a pack. "I could tell you weren't down with the brotha smokes, so I picked you up a pack of Camels. You look like a Camel sorta guy."

"Thanks," said Jeremy, then laughed. They pulled out their cigarettes and Jason lit them with his Zippo.

"I wouldn't take you for a menthol man," said Jeremy. "I'd think you'd want something a little stronger."

"Nah, I just like the way the box looks."

"I see."

"So, is your crib near Harlem?"

"Yeah. Relatively," said Jeremy, realizing that in theory his apartment at 92nd and Central Park West was in fact very near to Harlem, as far as geography goes.

Jason's question reminded Jeremy that his first trip to Harlem—real Harlem, where spirits of old prevail over the surroundings—was when some friends—"acquaintances," he would say, for he knew them only by sight and first names, if that—invited him up to their place for a party.

"We're having a party," one of the newly minted Harlemites had said when he saw me at a then-popular Downtown restaurant. "We just moved into this place in Harlem. You should come up if you get a chance."

Harlem? I thought, smiling. I smiled because I had heard this so many times before, from guys like these Columbia students who mistook 116th Street for Harlem, when in fact they were in some great apartment, financed by successful first-generation Columbia alumni. Had they been black, I might have thought maybe they did actually, possibly, live in Harlem. But these were white boys. Kansas white, submerged in everything that New York had to offer.

The evening before the party, they left a message on my machine, detailing the directions to their place.

"Take the A train to 125th and walk down to 118th. We live between Manhattan and Morningside," said the voice on the machine. I wrote it all down just in case I decided to go, and since I was unfamiliar with the street names, I was still apprehensive.

The night of the party, I decided to attend, but since I rarely went above 14th Street, I had to work really hard at motivating myself for the long journey uptown—the place they were calling Harlem. Still, I was in one of my wanting-to-meet-new-people moods, so I got dressed and walked to the L in order to connect to the A train at 8th Avenue. I found myself thinking of Ella Fitzgerald singing the Strayhorn tune about the very train on which I sat.

As the train headed north, I noticed the faces of the passengers grew darker at each stop. Gradually, the white faces disappeared, their places taken by a spectrum ranging from plum to yellow squash. On the platforms, I saw advertisements with beautiful, smiling brown faces, pushing Newport cigarettes and upcoming movies that someone in a Midtown office felt would be appropriate for the people in these neighborhoods.

At 125th Street, I got out to find cops and young men—black young men, wearing baggy pants and belts that knew not their purpose—walking around the station. I felt like I'd stepped into another world in this station. I could sense the difference, yet all this had become a normal existence to the train's frequent riders, cops and young men—black young men.

I took the stairs up to the street, finding the darkest dark imaginable. Even the moon hid. The only light was the vague glow that crept from the depths of the subway station, sprinkling drops of light to an otherwise bleak street. There were no street lamps to guide pedestrians, and no cars seem to exist. I became fascinated by the architecture, its beauty, its strength, its solidity and solitude. The vacant buildings were too old to speak, yet they spoke all the same. I passed a few people along the way. I felt comfortable here, not at all threatened by the new environment, and was amazed.

When I reached 118th Street, I looked left, then right, in a sense, a sleuth in search of a clue. I didn't know which way the building numbers ran, which made it difficult to decide exactly which way to turn. I walked around, looking like a customer in a video store with just too many movies from which to choose.

"Yo!" said a voice, but not cordially. It wasn't a greeting. The voice belonged to a guy who was one of a group of about five hanging out on the street. I didn't turn around, but out of the corners of my eyes, could see the guy coming toward me, the others following like the cars of the A train.

"Yo," the voice said again. Louder. I just kept walking, trying to ignore him and hoping that the guy wasn't speaking to me, hoping that would discourage him, hoping to just blend in with the buildings. But the guy kept speeding up. His footsteps becoming more aggressive, a race to the finish line.

"Wassup? You lost?" asked the guy, catching up with me. "You need somethin'? Cess. Ice."

"No. I'm cool," I said, still walking, trying to find the building.

"Then why you walkin' 'round like that, lookin' at buildin's and shit? You five-O?"

"Uh, no. I'm just looking for an address. Here it is."

The building appeared at just that moment, like lost keys right under my nose.

"You don't live 'round here, do you?"

"Uh, no. Just visiting friends."

"Well, if you need anything, just let me know. I'm around."

"Thanks." I knew I must have looked out of place in my psuedobohemian Downtown attire, strolling through their neighborhood, looking with amazement for the stories that once filled the now-empty beautiful buildings. The beauty of the façades wasn't enough for me, it was the spirit I was seeking.

To this group of guys, my having the same hue wasn't enough. They saw me as a stranger. A stranger in Harlem, a place I'd only read about but never been. I'd never thought that I, a black man, would be a minority here. Not in Harlem.

I rang the doorbell, and soon I descended from the dark street to the well-lit refurbished garden-level apartment. It was a beautiful place, no doubt of that. Much better than my little East Village studio.

I met the other partygoers, had drinks. There I was hobnobbing in Harlem and yet, again, just like on the streets, I was the minority. Though I hated to think in those terms, that's the way it was—but somehow, I was used to it.

The party had all the accoutrements: conversation, music, brie, newly invented mixed frozen drinks to quench the thirst. To the others, I wasn't a mixture unexplored. I wasn't Harlem, nor were my hosts. I wasn't the man behind the bulletproof window at the corner bodega who passed your purchase through the sliding window because the times had made him wary of letting anyone enter the store after dusk. No, I wasn't that. I was one of them and had every reason to be at this party with my friends, acquaintances, hearing Downtown stories in an Uptown venue.

"This place is so nice," said one of the guests. "It's almost worth moving up here."

"Yeah. I'm sure the real estate is affordable," said another. "You can probably get a brownstone for nothing."

"I wonder what it was like before the riots," said yet another.

"My father was in the riots when he was at Columbia . . ."

The conversation went on, but I had nothing to contribute. I too wondered what it was like. Kids playing, neighbors talking and greeting passersby. No constant cries of "Cess. Ice." No bulletproof windows symbolizing distrust. I began daydreaming about how it must have been before the great light of fire came through, leaving a trail of darkness. Now, it was no longer fire taking over these buildings.

When I glanced at my watch, it was 1:00 A.M. I thought of having the host call for a car service, for no yellow cabs were to be found in Harlem at night, but I decided I'd take the subway. I said my good-byes and thanked the hosts for a "delightful evening."

"Does the subway run regularly at this hour?" I inquired, as I always did when I was in an unfamiliar area.

Before I could get an answer, one of the guests jumped in. "Don't be ridiculous. The subway at this hour?"

"I don't mind," I said.

"We've got our car; you can ride down with us. We're going your way."

My way? *I thought, but thinking of the time I would save,
I agreed to the ride. So from the door to the car we went, then
zoomed down the streets of Uptown.*

*"They really have a great place," said one of the passengers.
"I'm really glad I came up."*

*"Me, too," I said, sitting quietly, taking in the view heading
back Downtown from Harlem.*

"So what's it like?" asked Jason, breaking into Jeremy's
thoughts. "A friend of mine was up there and said that you get
the freshest sneakers and gear on 125th."

"It's about more than just sneakers, Jason. It's beautiful. It's a
feeling. A vibe. I did a shoot outside the Apollo. Got everyone
great suits. I tried to recreate a long-gone era. It came out quite
well."

"Hey, boys," said Kim, heading across the lawn toward them.
She caught them off guard, and Jeremy jumped a bit. "Can I get
a light?"

"Yo, Kim. Long time no see," said Jason, jumping up to light
her cigarette. She was holding a baby, and Jason played with it.
Though Jeremy was "trying to quit," he couldn't bear to see a
mother smoking with her baby.

"You know where I live. But you just like all the rest."

"Nah, now . . . I been thinkin' 'bout you som'n fierce. But
you know, with the ol' man and all," said Jason, puffing on his
cigarette. It was as though he had been taking classes from Paul.
"You lookin' good, dough."

"Don't tell me; tell ya friends."

Jeremy started laughing, which made Jason remember that he
was witnessing his lack of progress.

"Have you met my big brotha—Jeremy?"

"We've—"

"Hey, Jeremy," she said, cutting him off, keeping their meeting
between them. "Wassup?"

173

"A preposition," said Jason, quickly claiming the joke as his own.

"What?" asked Kim.

"Nothing, Kim. He's just being *stoopid.*"

"Whateva. It's nice to meet you. I hope you ain't like this one here. He think's he's all that, but I know the truth. You know people get arrested for perpetratin'. And I have three words for you: Bourgie, bye-bye."

"Aw, baby, why you wanna play a playa in front of his peeps like that?"

"N. E. Way," she said, throwing her palm up to his face. "Is Jess in the house?"

"Yeah, she's in there," said Jeremy. "Nice meeting you, Kim."

"Uh-huh. You, too."

Kim went into the house, leaving Jason and Jeremy outside with their thoughts and respective cigarettes, wanting neither to ask nor to tell.

" 'Peeps?' I guess you picked that up in London as well."

"Come on, man. Give a brotha a break. You know how it is."

"Right."

"You wanna hear a joke?"

"I'm not much in a mood for a joke right now."

"Why you always gotta cut somebody off? Would it be so difficult for you to say yes sometimes? 'Yes, Jason, I'd love to hear a joke.' Easy as that. Damn."

"Sorry. I had a fight with Aunt Jess about . . . sure. Tell me a joke."

"Nah. See, it's too late now. You snooze, you lose, left to booze."

"All right. Fine."

"Okay. See, it's hella hot, right? And these three get-it girls — you know, get-it girls . . ."

"I've already heard this."

"What? Getthafuckout. They got that one up in New York?"

"No, Paul told me last night."

"I'm the one that told that nigga."

"And did you pick that word up while you were studying abroad with the Riverwood class?"

Jason said nothing. Jeremy felt as though he was again closing himself off from his brother and thought he should fill the silence.

"You know, the first time I saw you, that was the car he was driving," said Jeremy of Jason's BMW.

"Yeah, he gave it to me last year. It's awright, but it wasn't nearly as cool as your Rambler. That was da bomb."

"Yeah. You know, he gave me the old MG and I sold it to buy back the Rambler. Actually, I just traded it flat out."

"I can't believe the favored son did somethin' as whack as that."

"Yeah, well, believe it."

"What, were you on crack?"

This took Jeremy by surprise. Sixteen just wasn't what it used to be.

"You looking out for your sister?"

"She's your sister, too."

"Well, our sister?"

"She got hooked up with a new Jeep. The old man didn't want his little girl in a car that might break down on her. But it's all good."

"I didn't mean that."

"Yeah, I got her back."

"She writes me a lot."

"That's like her. She's pretty much like you, brainy and shit. But we get on. I have to keep the playas away, but she can handle herself. I think she thinks she's Cleopatra Jones," said Jason, throwing a karate chop through the air. This also surprised Jeremy.

"How do you know about Cleopatra Jones?"

"VCR, man. Hello?"

"Well, you seem like quite the player yourself."

"You know I get mine."

"I see."

"Don't get me wrong. I wear the glove."

" 'The glove'?"

"You can't possibly be as lame as you seem." But before Jeremy could respond, Jason continued. "Listen. Mom wanted me to come over and see you."

"Why?"

"She heard me last night. She don't miss a trick."

"Sorry about that. You were right."

"Oh, I know I was right. But I wasn't upset with you. I was just pissed that everyone had been talkin' about you and when you were comin'. Like you were the only one that mattered."

"Sorry about that, too."

"You shole is one sorry muthafucka."

"What's with the language?"

"Hey, you know."

"No. I don't."

Kim returned from the house. The baby had fallen asleep in her arms. They all looked at each other, but no words passed. She walked past them back to her house. Aunt Jess came to the door.

"Now, isn't that sweet. The Bishop boys sittin' out here together. Now, Jeremy, if I were a photographer, this would be the picture I'd take. Can't put a price on that."

"Wass—how are you, Aunt Jess, witcha fine self?" asked Jason.

"Boy, I told you about fresh-talkin' me like that," said Aunt Jess, blushing all the same.

"You know you like it. You are fine," said Jeremy. "You know, Jason, she refused to marry me because I was family—like that ever stopped anybody down here."

"Don't you try to go gettin' on my good side now, J. I'm still mad at you. I ain't forgot, so you can just save ya funny comments for somebody else."

"Aunt Jess, you know you can't stay angry with me," he said, taking her in his arms. She feigned resistance, then finally pushed him away.

"Y'all are both cut from the same cloth. If you weren't kin, I wouldn't trust either of you farther than I could kick you."

"Well, it's good we're kin then," said Jason, putting his arm around Jeremy. "'Cuz if we weren't, people would talk. Oh, I can hear Gloria now: 'You know, Jess over there keepin' comp'ny wit' two fine young mens. Brothas, no less, and . . .' "

"Lemmetellyasom'n," they all said together. The tension of earlier hours vanished, leaving their joy to fill the air, and the only thing Jeremy felt about his shoulders was his brother's arm.

chapter 13

⮌ ⮎

Cars came in droves during the course of the day, dropping off a smorgasbord of food, things Jeremy associated with New York delis. Three kinds of potato salad with different tints. Jell-O with fruit stuck in it. Macaroni salad. Green-bean casserole. Chicken wings. Rib, ribs, and more ribs.

Jason sat out in the yard with Jeremy as he watched car after car of don't-you-remember-me faces speaking with goading familiarity. The house and yard spilled over with levity, making the heaviness of the occasion that brought everyone there waft away like the scent of honeysuckle after rain.

The younger people stayed outside while the older went in, each commenting on the burdens of the other. Funerals were different from weddings. All less than favorable details were put aside when it came to dealing with the end; the future barred resentment and judgments, for the dead are no longer a threat.

Jeremy still felt out of sorts. Everyone had so many good words to say of his father, and the stories and knee-slapping jokes filled the house. He wished that he could think of something, but it was evident that they knew him better than he did. They seemed to know a different man.

He attempted to imagine if this was how it was when his mother died. No one had ever spoken of it. Was that to protect him, or was in an attempt to move on? He couldn't see how any laughter could have filled the rooms—the very cause of her death had still been in the house, lying on his back in delight at the new world, completely unaware. He'd never even been to his mother's grave. He had never asked of its whereabouts, and no one ever offered to tell him.

Jeremy had become a photographer so that he could capture faces to replace the one he'd never seen. Everyone assumed he'd be some sort of writer. He had been forever writing stories and poems, reciting them at will, but when adults would bend over him, asking what he was going to be when he grew up, he'd say, "A photographer." This always brought bewilderment, and the grown-ups would straighten up, then arch backwards, wrinkling their noses as though his words carried a stench. *Photographer* wasn't something one would expect to hear from a young boy, at least not a young black boy. *Fireman, lawyer, doctor,* fine; even *garbage man* would generate a grin or two, but *photographer* never captured the imagination or gained him applause.

That's nice. We all need a hobby, but what do you want to do for a living?

I don't know no black photographers.

Now, what you should do is follow in your daddy's footsteps. A dentist—that's a fine profession.

Yes, be an orthodony. That's the ticket, there.

Now, with outstretched hands, those very people congratulated him on his accomplishments. Jess's phone was like a newsletter: she told everyone of any of Jeremy's prints that appeared. When one of his shots of Bill Cosby appeared on the cover of *TV Guide*, they knew it was true that he had made it.

For Jeremy, it was just a job, another photo. His work didn't change the world with its social commentary; it wasn't hanging on a SoHo gallery wall or gracing the pages of *Life* magazine. But he was at the top of his field and good at what he did, and that satisfaction had eventually become enough. His success had often made him feel guilty. Black men weren't supposed to be successful, and in the confines of the white environments that his job made him frequent, he was surely reminded of that.

He thanked his father's friends for their support and words of encouragement, refusing to show any animosity. They now knew a black photographer and the road for others would then be less arduous, and that was respectable.

The house had finally emptied. Jess had retired to the rocker. The kitchen and dining room were covered with different pots and Tupperware containers, all purchased just for this occasion so the mourners wouldn't have to be troubled with returning them to their proper addresses. Jeremy had been picking on food all day, keeping himself from having to partake in too many conversations. He wanted to be more forthcoming, but he was overwhelmed by crowds. He had repeated the same answers for the same questions concerning his life in New York so many times that it became surreal.

"It's been a long day," said Jess.

"Yeah, it has."

"You goin' out tonight?"

"Maybe. I don't know. The days seem to last longer down here than at home."

"Well, we have two speeds down here: slow and mildew. But I'd like to think that this is your home."

"You know what I mean."

"Now, just 'cause I'm talkin' to you, don't think I've forgotten about this mornin'."

Jeremy chuckled "No. I didn't imagine that you had."

"You know you were showin' out. We raised you better than that. Yes, you went through a lot. I know. Who hasn't? There's been more bad draped over you than someone ya age should have to know, but you still here."

"Aunt Jess, you can't expect me to feel more than I feel."

"You're right. But how long can you deny ya feelin's? Yes, ya daddy may have made some bad choices; we all have. There were some times I should have knocked you upside ya knucklehead, but I didn't. But how long are you gonna carry those bad feelin's around with you?"

"Why is it that everyone always talks about him, but never my mother?" A pause filled the moment like a drag on a cigarette. Jeremy had been waiting to heave this into the conversation, waiting to force it to shift the focus away from himself and make some sense out of what might have been. "Things would have been different had my mother lived."

"Coulda, woulda, shoulda. Who's to say? Nobody oughta live they life like that."

"It's like everyone wants to remind me of how great my father was, but what about my mother? It's as though she didn't exist."

Jess looked at Jeremy for a long time, as if he had slapped her and the sting placed her in shock. As far as she was concerned, he had been raised as though he were a child with a mother, a shield from harm's way.

"You sound like you disappointed with how you came out. We tried to do our best . . ."

"You know that's not what I meant."

"Then what do you mean?"

"I just find it peculiar that after all these years, no one even speaks of her. Every year, you send me a birthday card and you call, and you know that every birthday reminds me of her, but you never mention it."

"We just thought that you didn't want it mentioned. You never asked, and as time went on, the less we thought of bringin' it up. Why open old wounds?"

"I'm asking now. Wounds can heal, but you have to treat them first."

Again, an oblique silence filled the room. It was so quiet that the tick of the electric clock sounded an outburst.

"Fair enough," she said. "I don't think this is the best time for this, but if you so set on it . . ."

"I am."

"You just can't let well enough alone, can you?"

"Because I'm not well enough."

"Awright then," she said, scratching her eyebrow. "Your father met her in college. Like you, your father couldn't wait to get away. He never got along with Papa, for reasons neither of them could control. Men can be stubborn when things ain't the way they want them to be. Of course, women can be just as stubborn. He met Helen, they courted for a while, and she got pregnant. Your father loved that girl."

"Did you love her?"

"Well, I was protective of your father. He was my little brother and had gone through a lot. But I grew to love her."

"How so?"

"Because of you."

"And my mother's parents?"

"They weren't at all pleased. See, ya mother had weak blood. The doctors said she wasn't supposed to get pregnant 'cause it would be hard on her. Your father didn't know that at the time. After she was pregnant, they got married—not just 'cause they had to; they wanted to. It had nothin' at all to do with the baby. That just pushed things up a bit.

"Her parents didn't take well to either news. All of the doctors, specialists, and the lot told her she probably wouldn't carry the baby to term." Jess stopped for a moment, then continued. "They recommended terminatin' the pregnancy."

"But I was born in '70. Abortion wasn't even legal." Jess looked at him with a sad stare, and he wanted to take the words back, fully aware they were just naiveté, grasping. Jeremy looked down at the floor. No matter how she phrased them or attempted to distance him from their meaning, the words she had been saying weren't just words—they were him, making past and present one. He stood up but went nowhere. He sat back down.

"You sure you wanna hear this?"

"Yes, please," he said, flicking the butt of the cigarette, not so much to rid it of ash, but to release the energy charging through him like electricity on a humid day. For a moment, he wished that his mother had had the abortion; then she would still be alive. But for her to be alive . . .

"Ya father wanted her to end it. Truth be told, we all did. But her parents were against that, too, and so was she, but for different reasons. Helen knew the risks and accepted them. I think she wanted to show her parents that they were wrong about ya father and that all would work out accordin'ly once the baby arrived. Love is a mighty powerful thing. Lots of fights went on. But nothin' was gonna stop her from havin' that baby."

It amazed Jeremy how far removed it all seemed from him, a grown man sitting listening to the story. It was as though he weren't real at all. It was a story of someone else, not him.

"They moved in there."

Jess motioned toward the room that he was raised in. He had shared the same bed as his mother and father and never realized that he had been so close to both of them, together.

He placed the half-smoked cigarette in the ashtray. The smoke rose, filling his vision. He picked it up, and though he dabbed and pressed at the end of it, the embers refused to fade and the smoke continued to rise. Finally, he thumbed out the spark. The second of burn that he knew he would feel didn't at all deter him from doing it, for pain rarely lingered far from his heart.

"Her parents never called. Not once. She acted like it didn't bother her, but it did. Tore her up inside. They had no more daughter as far as they were concerned. She had disgraced their family by getting pregnant out of wedlock, disgraced them by being involved with your father, who they felt wasn't good enough for her. They were a prominent family in Shreveport and thought everyone over here was just hillbillies or somethin. It's easy to not like what you don't know. But if they could have seen the two of them together . . . she brought happiness to his face, and this house needed that.

"One night, Ma Dear had had enough of the foolishness and called her parents. They thought we were callin' for money, and that made Ma Dear's blood rise. She was a proud woman, too.

She had already paid for everythin' without a thought to it—the doctors, the blood work, everythin'. She told them that she just wanted them to know how things were going and that the baby was soon due and that if they tried, she was sure we could patch things up. They told her not to call back again. As far as they were concerned, she was our problem. Wasn't a problem at all, really. But no one wants to feel that their parents have cut them off."

Jeremy grunted and rolled his eyes away from Jess. She refused to acknowledge him, refused to stop the telling. He had wanted to hear it, and now he was going to.

"One night, she started havin' stomach pains, but not contractions. It was a bit early yet. She was hurtin' so bad and bleedin' that we didn't even drive her to the hospital. The ambulance came to us. She tried to tell us all that it was goin' to be just fine. When she got there, she had a C-section because the placenta was blockin' the passage. Your daddy wouldn't even go in the hospital because he was too nervous. Men didn't go in the delivery room back then like they do now.

"I stood outside of the hospital with him, and he prayed. I hadn't seen your father pray in God knows how long, but he got right down on the ground."

Jeremy wondered if he prayed *only* for her and then hated himself for the selfish thought. But he wondered nonetheless.

"Word came out that the baby was a healthy boy and she was stable, but they were keepin' her in ICU. Your daddy asked if he could see her, and they told him that she was still under, but he could go in for a minute. I walked him into the hospital. He was like a child and an old man in one, barely able to find his footin'. When we walked in, Helen's mother was in the lobby by herself. Ma Dear had called her again, and the driver drove her from Shreveport. She told Ma Dear her husband refused to come. She looked at your daddy and told him to take care of her baby. That was all. She didn't go up to see her daughter, that I know. She said what she had to say, then she left. Your daddy went up to ICU, but when he got to her room, he wasn't allowed in."

"Did he go look at me?"

Jess didn't look at Jeremy. She couldn't. She had loved this boy more than anything and she knew his pain and didn't want to contribute to it, but she had to answer. She turned to him and he met her gaze, and in that lay his answer.

He took his eyes off her and took out another cigarette. He could barely hold it—his hands felt as if winter had found them gloveless. He finally got the cigarette lit, and she continued.

"He went back outside. He needed air. He felt he was to blame for her bein' there, and he carried that with him like a cross to bear. He kept sayin' that he should have made her . . ." She paused, as she knew what she was saying. "Well, I told him that he wasn't responsible, that it was in God's hands. He said that God wasn't the one up there in that bed, so somebody oughta wake God up. Ma Dear slapped him. She had never laid a hand on him before, but she did then.

"He told us he wanted to be alone, so we left him and went to wait in the lobby. A while later, though it seemed like forever, the nurse came down to say that she was still groggy, but she was conscious and the baby had been brought in to her. I wanted to run out and get your daddy, but I wanted to see for myself that she was all right first. I needed to know that he wasn't going to be slapped twice in one night.

"I peeped in on her and the baby was there in her arms. She had just the smallest smile on her face and her eyes were closed. Even after all she had gone through, she looked like an angel holdin' the baby. I thought she was restin'. I walked over to the bed. I called out her name, but . . . but she didn't open her eyes. I touched her head to wipe the hair from her face and . . ."

Jess stopped. She had been directing her view anywhere but upon Jeremy, but she turned right to him. "You were the last thing she saw, and maybe that's why she was smilin'. Your father didn't let anyone come to the funeral. He just had her buried.

We begged him not to go through that alone, but he did it anyway. He wasn't himself. You could see it in his eyes.

"With all the craziness going on, he hadn't named you, so Ma Dear started callin' you Patience. That's how that name truly came to be. It was also Ma Dear that decided your proper name, Jeremiah, and that was what they put on your birth certificate."

Jeremy had not taken a drag on the cigarette, and he had been so still that its ash grew to a cinematic length. When he did finally move, the ash fell to the floor, collapsing on impact.

chapter 14

～✑✑

He drove because it was what he wanted to do. He had no forethought as to where he was heading, but he ended up in the hospital parking lot. He sat looking at St. Francis, the old façade and its new additions, wondering which room it had been, which room had held the angel's smile.

He didn't stay long, he just wanted to drive.

Football stadium.

Exit that led to Crossing Pointe.

Cemetery where the third tree used to be.

Father's house.

And finally he stopped. The sign read JOHNSON'S FUNERAL HOME. Inside was where his father's body rested. He got out of the car, then sat on the hood, his weight making the hood sink in. He thought of the dent in the MG he had made years ago—and the one it had made in him when it was sent from California. He thought of his father and his mother and his mother's parents, all the people he never knew. It all came to this, and he'd been here before.

The airport lounge was crowded. Thursday night always meant the college students were out having a big hoorah before heading to their hometowns for the weekend. Jeremy walked to the empty table in the corner and claimed it as his own. He watched the planes come and go, arrive and depart. His "Johnnie Walker

Black, neat with a water back" sat with him, along with the
sounds of the Super Pac-Man and jukebox.

"Hey, you know it's not healthy to drink alone." It was Paul.
"At least not in public."

"Don't tell me you still hang out here."

The two had come here often to listen to bad music, watch
the planes, and flirt with anything that looked like someone waiting for a connection. They never ran into any of their peers here;
the scene was too common, and it was that that appealed to them.

"Sometimes," said Paul. "I just felt like it tonight. Beth was
trippin', and when a woman is trippin', that's one thing—you can
just tip around her and lay low. But when she's pregnant and
trippin', you gotta leave the premises."

Paul had hoped for a mutual laugh, but it didn't happen. He
couldn't have possibly known that any thought of a pregnant
woman, cranky or otherwise, was far from funny to Jeremy at
that moment.

"You look like shit. Not just shit, but shit flattened by a tire,
and all that."

"Thanks," said Jeremy. Still no laughter. Jeremy took a sip from
his scotch, but the water back remained untouched.

"Listen, man. Your Aunt Jess called me. She sounded worried.
You cool?"

"Damn, what am I—one of your foster kids or something?"

"Well, you actin' like it. Like you the only one in the world
and can't nobody do a damn thing for you, 'cuz you just know
you gonna get screwed over again. That's fine for kids, but you
ain't been a kid in a long time. Come on—it's me here."

"Is it?"

With Jeremy's response, he motioned for the cocktail waitress
and ordered a pitcher of beer. "And keep that big hair out of it.
I oughta call the Environmental Protection Agency 'cuz I think
I've discovered the problem with the ozone layer."

Jeremy relented with a laugh. He didn't want to, but he
couldn't contain himself. Not this time.

"You're crazy, you know that?"

"Ah. It's alive," said Paul. "It's alive."

Jeremy let out a moan and eased back into the chair. He rubbed his hands up and down over the stubble that massaged his palms and the faint sound of it reached his ears. Paul didn't attempt to coerce him into conversation. He knew that he'd broken the barrier and that in time, when Jeremy was ready, words would follow.

The waitress brought over the pitcher and two glasses.

"Thank you," said Jeremy by force of habit, but then he turned back to Paul. "I feel like a hypocrite being down here. This funeral business. I mean, I never really knew my father. Never tried to. Then Aunt Jess tells me this story today about my mother and—"

"Yo, ya dad loved you. Every time I saw him, all he ever talked about was you. He'd call you 'my boy.' 'Have you heard from my boy?' Hell, he talked about you to my pops so much, I think my ol' man was ready to trade me in for you."

"Well, he did a pretty good job of keeping it from me."

"Yeah, and Pat never told me he was gonna slam a bottle of pills, but he did. And even if he had come to me, what would I have done? Probably pushed him to the curb, told him to tell it walkin' and stop trippin'."

Paul took a drink from his glass of beer. As if out of nowhere, Jason appeared, grabbing the glass from Paul's mouth and taking it for himself. "Can I get another glass over here?" he screamed to the waitress as he pulled up a chair.

"Can I get an ID?"

"Ah, now, don't be like that. I'm more than old enough," said Jason, licking his lips.

The waitress raised her eyebrows and Jason smiled at her in his most charming way before throwing in a wink.

"He's my brother. I'll vouch for him."

She walked back over to the bar and brought back a glass.

"Well, havin' a big brother has finally paid off."

"Yeah, well, don't make a habit of it and don't tell Carol."

"So wassup witcha goin' 'round tryin' to get props for my joke?" asked Jason of Paul. "I'ma have to have my peeps talk to yo' peeps, but I'll settle out of court if the offer proves to be sufficient."

"That's right. I thought you said one of the kids told you that joke," said Jeremy.

Jason looked at Paul, and Jeremy, at both of them.

"Well, I do some volunteer work with Paul."

"What? Mr. Cool is doing volunteer work?"

"Get off me, I-ight? It's just community service for a little misunderstandin' I got in. This dude got all up in my face, talkin' shit, so I had to take him out. You gotta show muthafuckas just 'cuz they carry they food stamps in a money clip don't mean they all that."

They quickly put away three pitchers of beer and a few shots of tequila among them, and their conversation passed as rapidly. Peanut shells covered the table and the ashtray was filled beyond capacity. No single thought held sway for long. That corner table became a room of its own.

"You know, I never used to smoke, then I moved to New York and the next thing I knew, a pack a day. It's like it's a part of the persona that you develop because you become something new. You re-create yourself. All indications of innocence have to be hidden because people are just waiting to pounce on anything that shows weakness. But once you lose your innocence, all that's left is fear, angst, and cigarettes—and a person in the mirror that you don't even recognize anymore. It's just a semblance of you. And there's no one to blame. You know?" said Jeremy.

"You remember when we used to ride around, and every time we ended up at a red light next to some uptight-lookin' white woman, you would look over at her real scared like and push down the lock?" asked Paul.

"I went to the movies, right? I get up to the window and I say to the clerk, 'Yeah, give me a ticket for so and so,' and the woman said she was sorry but that movie wahn't showin' there, and I said, 'Damn, the commercial said, "Now showing in theaters everywhere," ' " said Jason.

"Dat's my Leroy," said Paul.

"When I die and if the two of you are still alive, promise me that if you ever see carnations on my grave, you'll take them off," said Jeremy.

"No more wire hangers!" they all shouted, and the remaining patrons looked concerned, which only made their outbursts escalate.

The last flight had landed. They had closed down the bar, but they were the only ones this seemed to impress.

"Well, boys, there's a wake tomorrow, so we oughta be takin' it to the house," said Jeremy.

"Where's Christmas when you need it?" asked Paul, pulling out his cellular phone.

"Whatchu talkin' 'bout?" asked Jason.

"Tipsy Taxi, muthafucka."

They sat on the curb outside the airport, all too drunk to drive, almost too drunk to speak anymore. Every now and then, one of them would laugh without acknowledging why and the other two would just join in as though they felt the same. Carol drove up, and they all stumbled into the back of the ever-practical minivan. She had on her robe and rollers circled her head.

"You all smell like a brewery, and I hope you throw up," was all she said, and even then, they couldn't contain their laughter.

*   *   *

When Jeremy opened his eyes, he was lost. His sleep had been sound, like the sleep of being in the comfort of ones own bed after a brief absence, but when he awoke, he didn't know where he was. This wasn't his apartment or his Aunt Jess's. His mind retraced his steps. It took him a bit to focus and realize that he was in his father's house. This was the room he'd lived in for the years sixteen through eighteen. It looked exactly as it did back then—merely a guest room. No posters on the walls. No pictures of him or anything that said he lived there. No additional personality inhabited the hotel-like room.

He was wearing only his underwear but couldn't remember undressing. He couldn't remember driving here. All he could remember was sitting in the airport lounge and saying something about the loss of innocence. Everything else after that was fragmented. Then his conversation with Jess added to his pounding head. "Oh, God," he moaned, putting his hands over his face as though none of it had really occurred and it was all a dream, a nightmare.

Here he was in this room again, forever with a closed door. He would leave in the mornings and return late at night when he knew everyone was asleep. He kept his door shut and his seat at the dinner table often remained empty.

I was walking lightly through the house late one night, as I had begun to do regularly. I was an expert at navigating my way without a hint of light. As I passed through the den, he was sitting on the sofa in the dark, waiting. The light's being switched on and his voice both startled me.

"We need to talk," he said.

"I'm tired. I'm going to bed," I said, walking on.

"I'm tired, too. I'm tired of your acting as though you're not a part of this family and punishing everyone in this house. You're not the only one who lives here, and it's about time you acknowledge that."

"Fine. I'll go live with Aunt Jess."

Our voices remained just above a whisper, voices of private confrontation. Though we tried to avoid conflict with one another, that very avoidance made me want to instigate it; fighting came to fill the void of concern and affection.

"You're such an ungrateful little bastard," he said, then looked almost as though he hadn't wanted the words to leave his mouth. That last word trailed off. I looked at him, smiled, then started toward my room. "Don't walk away from me."

Again, I looked at him, but rather than smiling, I laughed. I couldn't seem to stop, for the words kept tickling me. "Do you hear what you're saying? You're the one who made me a bastard, and you're the one who walks away. You think I'm supposed to jump for joy because good ol' Dad decided to let number-one son come live in the Big House with him. Well, I've done just fine without you."

As testosterone took over, our voices were no longer whispers; they were lions' roars.

"You have all the answers, don't you? One day, I'm gonna knock that chip off your shoulder."

"I'd like to see you try," I said, almost challenging him. My fist were clenched, waiting to unleash the years of resentment coiled in my spine. But it was he who laughed then. As if on cue, Carol came into the room. It was as though she had a sixth sense when it came to conflict.

"What's going on?"

"Nothing, Carol," he said, still laughing. "This is between the two of us. Go back to bed."

"Yeah, 'nothing' is exactly right. I just caught some imposter trying to play father." With that, I left them standing there. I went to my room, closing the door behind me.

Yes, yesterday seemed like a bad dream. He got up from the bed and went into the bathroom. Jason and Jessica shared the bathroom that divided their bedroom, but he had a private bath in his room. "A young man needs his privacy," his father had said when Jeremy moved in.

He washed his face and looked through the drawers and cabinets for aspirin. His movements were slow, mechanical, and his ears were still ringing with the roars of planes accelerating into the beyond and the chomping of the ever-hungry Pac-Man. He thought a shower would help. He turned on the water, then pulled off his underwear. He looked at his body in the mirror. The words *terminate*, *C-section*, and *the baby* all seemed to be scrawled on the mirror, making it difficult for him to see himself. He stepped into the shower. As the water ran down his face, he imagined his mother's smiling face. He soaped up the sea sponges and scrubbed hard and long. He wanted to be cleansed from it all—the hate, the loss, the memories. He scrubbed until exhausted, his muscles sore, his skin irritated.

When he was dressed he walked out of the room. Just like old times, his steps were tentative, as if he expected a confrontation.

"Good afternoon, sleepyhead."

"Hey, Jessica. My God, you've grown," he said, giving her a slight hug, almost as though she were too old to embrace.

"You sound like an old person. That's all they ever say. If you say 'If you do everything now, what will you have to look forward to when you get older?' I'm going to swear you aren't my brother."

"No. Sorry. It's just that it's been a while. You've just . . . just."

Jessica had developed into a young woman. Though she and Jason were twins, Jeremy hadn't really noticed Jason's development—Jason seemed the same to him, but Jessica's transformation was evident.

"Please don't give her a big head," said Jason, walking through the living room in nothing but his boxer shorts. They were red, and were covered with different tools and their names: jackhammer, screwdriver, and pile driver. "She already thinks she's Miss Louisiana."

"Ignore him. I'm convinced that he was dropped on his head

as a baby. That's why he grew those dusty dreads, so he could cover up the scar," said Jessica, punching Jason's arm with all her might. "So I hear y'all really tied one on last night. You know, drinking isn't going to make this go away."

"Well, thank you, Dear Black Abby," yelled Jason from the kitchen. "Jeremy, I told her she watches too many talk shows."

"Whatever. And go put on some clothes. Nobody wants to look at your ashy body," she said, then adding, "It's a bird . . . it's a plane . . . no, it's Ashman, coming to black skin near you."

The smile on Jeremy's face found him torn. He watched how the two interacted seamlessly, throwing insults that weren't at all meant to harm. They had a relationship that went further than their being twins. They were family. As he watched them, it pained him. He thought of how he had always considered himself an only child. He had never known this camaraderie that he was witnessing.

Carol joined them in the kitchen. They had breakfast in the room off the kitchen's bay window, the chairs around the table all filled. There used to be five chairs, but when Jeremy had left for New York, the fifth one was removed. Now the seat he filled was his father's, and he could sense everyone's awareness of that as they all sat in their places.

Carol told him that after breakfast he should call Jess. She had phoned earlier to tell her he was fine, but Jess was still concerned about him. Carol looked at Jeremy for any clue as to what he was thinking. Jess had told her that he'd been told the story of his mother.

After Mama B's funeral, Jeremy's father had mentioned to Carol that Jeremy was called Patience. That had evidently lodged in her mind. His first day in the house, she tried to call him that. He showed no patience in making it quite clear to her that she couldn't.

"You know, Patience, we're all happy to have you here," she had said, trying to reassure him about his move.

"Don't you ever call me that," he said, spitting each of the words at her. "Just because I'm here doesn't mean I want to be, and if you ever call me that again, I'm outta here." He had left her standing there, so he didn't witness the tears that followed his reprimand. She didn't tell his father, for she was trying to ease the tension that was part of any transition, and she and Jeremy never spoke of the incident again.

Jeremy did warm to Carol eventually, and in his own way, he grew to love her. She was a good woman, but she was his father's wife, and that was enough to make him keep his affection a secret. She had always been there, apologizing for his father, trying to justify his actions. He never understood how anyone could possibly love his father, and he never wanted to understand.

As he looked across the table at Carol now, he thought of his own mother, and if *she* had loved him at one time, perhaps someone else could as well; perhaps there was hope for him.

The topic of the wake took front and center, making everyone pick at the crust of their marmalade-covered toast, instantly unappetizing. All heads but Carol's were down as though in prayer. She told Jason to be sure to pick up the clothes from the cleaners before three and Jessica to pick up the thank-you cards from the printers. She reeled off these things as calmly as if they were the daily chores. When the instructions had been issued, she picked up her plate of food, barely touched, and excused herself from the table.

The other three remained seated, appetites lost. The over-easy eggs on Jeremy's plate took him back to the previous night, and his stomach told him it would rebel at food. The *Elsewhere World* remained in its rubber band in the middle of the table, untouched. The news of the day seemed unimportant.

"Hello."
"Jeremy?" said Jeremy, voice disguised.
"No, this is Jess. Jeremy's not here right . . . J? J, is that you?"

"Yes," said Jeremy, laughing into the phone.

"You know you crazy. You know that, don't you?" Jess's voice conveyed her relief that he sounded in good spirits. "Heard y'all were actin' like it was the night before Prohibition last night."

"Yeah, we did close the place, sad to say."

"Are you awright?" They both knew the question had little to do with the hangover that bit into him like a pit bull.

"Yeah. I'm fine. Thanks."

The thanks wasn't so much for her asking; rather, it was for what she had already told him. Now that he was no longer covered in it, he could appreciate the story that she had relayed.

"Hold on. Someone wants to talk to you," said Jess.

"Jeremy? Baby, you there?"

"Gloria, is that you?"

"You know it is. What other gal you got down here callin' you 'baby'? How ya doin', sugah?"

"I'm fine."

"Now I didn't say nothin' 'bout how ya lookin'—I ast ya how ya doin'."

"I'm well," Jeremy said with a laugh, his head no longer pounding enough to contradict that statement.

"Betta than that, from what I hear. Listen, you can make a check out to me, and use as many zeros as you want, hear, baby?"

"You're not gonna have all those men of yours trying to kill me?"

"Lemmetellyasom'n."

chapter 15

∽ ⌒

"Nice Jeep," said Jeremy, walking out to the garage.

"Yeah. It's okay. A sexist gesture, really, but it gets me where I'm going," said Jessica, sliding behind the wheel with ease. "I'd rather drive the Beemer and Mophead likes the Jeep, so we trade off."

Jeremy couldn't help but notice her attire—boys' boxers and a T-shirt tied in front that exposed her midriff. "They let you out of the house dressed like that?"

"Oh, puh-lease."

Jeremy settled into the passenger's seat, his seat belt so tight it pinched his chest. Jessica drove the same way she spoke, aggressively. She seemed effervescent and fearless. Aside from Aunt Jess, Jessica had been the only one from home with whom he had frequent contact. She'd sent cards for the holidays and wrote extensive letters detailing the goings-on in her life.

"I'm a lesbian, you know." she said, turning a sharp corner.

"What?"

"Not really. Just checking to make sure you were still with me."

"I mean . . . it wouldn't matter to me. I mean—if you were," he said, attempting to play the understanding older brother, something with which he had very little experience.

"Don't be ridiculous. I may explore when I go to college in a couple of years, but I'm not a lesbian. I'm a liberal. I'll try almost anything once. Sarabeth's sister went down to Tulane, and the next thing you know, she told her parents she was a lesbian, got them all in a tizzy. But right after graduation, she married a guy who's a lawyer. Turned out she was just a LUG."

"Lug?" inquired Jeremy, completely taken off guard by the conversation.

"LUG. Lesbian until graduation. God! Where have you been? For someone who lives in New York, you're pretty out of the loop."

"Sorry to disappoint you. I'm just not up on the current lesbian lingo. I'll make certain to buy the latest sapphic dictionary when I get back."

"Don't worry about it. No matter what, I still love you."

Jeremy felt as though she had somehow reversed the roles; she was the adult and he was the teenager. Oddly enough, he did feel she loved him.

"You're sixteen, right?"

"Seventeen, this summer. Send cash. I'll give you the direct-deposit info if that'll make it easier."

"I'll keep that in mind."

"Do that. Now, listen," she said, adjusting her sunglasses while glancing in the rearview mirror. "Mom's being really strong about this death business, but it's eating her up. They had been fighting off and on."

"About what?" asked Jeremy, taking a pair of Ray-Ban navigators out of her ashtray, which she used for accessories of a different sort, none of which were cigarettes.

"You can have those if you want them. They were Danny Zimmer's. Captain of the varsity baseball team. I was going out with him for a while. Well, not really a while, just for the homecoming dance. Long enough to get his letter jacket. He was good-looking, I mean truly fine, but I had to give him his walking papers. See, you men somehow believe that if you don't hit a home run your first time at the plate, then you're gonna be sent back down to the minors. Well, it works differently in my organization. I'm the owner of this team, and I'll give you a chance to prove yourself. But patience is a virtue, and you won't be stealing any bases here."

Jessica pulled the Jeep into the parking lot of Judy's Paperie. She slid out of the seat, waving to people she knew, or at least people who seemed to know her, and was at the door before he was even able to unstrap the seat belt. Rather than walk, she seemed to float. Her self-assured demeanor and height had garnered her the center of the Riverwood High kick line. She was like a butterfly in spring, bouncing and swaying in its beauty for all to see but not really concerned with whether anyone noticed, for it was certain that you did.

"Just stay in the car," she yelled over her shoulder. "I'll be right out."

Jeremy had just opened the door and put his leg out, but he returned to the seat and waited. He was under her spell. He had been told she was a lesbian, then wasn't a lesbian, forewarned about Carol, given sunglasses of an ex-boyfriend and a summation of men that entailed a baseball metaphor, but about the . . .

"They started fighting regularly after you left," she said, getting back into her seat and throwing the box in the back. "You buckled up tight?"

"Yes."

"Carol ragged on him about how he hadn't made an effort to be a good father to you and that it was inexcusable. I love that word. It sounds so delicious off the tongue: *inexcusable*. It's just the most perfect word, really. Don't you think?" Without waiting for his agreement, she continued. "See, Mom felt that you thought she was standing in the way of y'all's relationship. That little 'discussion' went on for a few months. Blah, blah, blah—or blah, cubed. Don't you like that—'blah, cubed'?"

Again, she allowed him no answer. "That went on for a while, then it cooled a bit. Mom had a big case, Dad was busy, la di da di da. Then the glitter really hit the ceiling. See, every Sunday morning, Dad would just disappear. I mean, you know, he was always busy with the practice, always working, sitting on this board or that committee, but these Sunday outings, well . . . Mom got all out of sorts and was convinced that he was fooling around. Midlife crisis and all. She was certain that he was seeing his assistant at the office. *She's* okay, but I wouldn't think she would be his type. But who knows? Better have done worse. But let's stick with content; we'll deal with mess later."

She turned down the unrecognizable song on the radio and looked at Jeremy as though he were a child. She put her right hand on his leg. "This is strictly between us because I shouldn't even know this. But what's a girl to do with her extra time? Somebody's gotta look out for the well-being of this family."

"So was he having an affair?"

"We don't know. He kept saying that he was at the golf course. *Lies!* Because one day, Mom sees Paul's dad at the Quik-Stop, and they are—were—golf partners. Carol made a beeline for the gas pump, but the tank was full. She got out of the car, and in that lawyerly way of hers, she asked Mr. Daley if they had canceled for the day, and he said that tee time wasn't until noon, which is silly, for anybody in the know knows tea time isn't until after three."

Jessica laughed at her own wit, but Jeremy just wanted to hear the outcome of the story.

"So there we were, wondering—well, Mom was wondering, because of course, I was still playing clueless—what was going on in those morning hours before noon. She wanted to wonder no more, and so she confronted him about it. She said she was not going to be made a fool of. No—a 'laughingstock' is what she said. Such an antiquated term. I'd really expected more than that from Mom. *Laughingstock.* Not nearly as pungent as *inexcusable.* But being fearful of the doghouse, he swore up and down that he just took those hours to be by himself. Now, the plot thickens. Around this time, we started getting a lot of hang-up calls. Mom wasn't having it. She got fierce and got Caller ID. As it turned out, most of the hang-ups were for Mophead."

*   *   *

While they ran errands and Jeremy feared for his life because of Jessica's driving, he was informed that Carol had been sleeping on the sofa in her bedroom since her husband died. She couldn't sleep in the bed, because that's where she found him. By the time he had informed them that he was sick, it was too late. He didn't want to go to the hospital. His appearance had begun to change, but he had been on a special diet because of his blood pressure, so the loss of weight wasn't at all alarming; he'd even been complimented on it.

He had been bedridden for the last few weeks. He made Carol promise that she wouldn't tell anyone. He didn't want to be seen when he was weak. She had taken leave from her firm to stay at home with him. He had been in good spirits until the end. She went to the kitchen to get him something to drink, and when she returned, the smile on his face didn't let her know right away that he was gone. Jessica had found her in their bedroom, sitting on the sofa. She had been there for hours. She hadn't called anyone; she just sat there, looking at his smile. She had cried, had laughed, had felt angry at his concealing this from them— from her—for so long, and if he would keep something like this all bottled up, then what more hadn't he told her?

The linens had been changed and the bed made up as beautifully as a window display, wonderful to look at, but never slept in. Carol was a strong woman, but law school had not prepared her for this.

I had the opportunity to go to the university of my choice. Once I realized that it was all right to do so, I had excelled in school. But I decided that the school environment was not suited to me, which is easier to say when you're given options.

My father was angry when I told him that I wasn't going to Hampton or Howard or Berkeley. He had written detailed letters on my behalf to associates of his who could supply me with the appropriate recommendations. I wasn't at all interested in these big-name schools, for I had learned rather early in life that the name doesn't make the person; the person makes the name.

When I received the acceptance letters from these institutions as well as a few other Ivy-covered buildings and abstained from replying, all he said was, "All I've done for you is going to be wasted. If you want to ruin your life, that's fine, but don't expect me to support you while you 'find yourself.'"

But even he knew those were futile words. I'd never felt that he had supported me. During the arrangements for Mama B's funeral, my father and his new family had closed on a house and moved back to Elsewhere shortly thereafter. I was summoned to live with them in what they would call Bishop House. The house might have carried my name, but I can't say that it felt like home.

He didn't even ask me or talk to me about it. He had decided that it would be better this way. He had Aunt Jess break the news to me.

"It's time you get to know ya father," she said.

"I think I know pretty much all I want to know."

"Nobody does, J, and if you think you do, then you not as smart as I give you credit for. Why don't you just give it a try."

"But what about you?"

"Don't worry about me. Besides, it's just across town. It's not like you leavin'. I can see you every day—not that I want to," she said, breaking the tension with a joke.

And that was true, but those few miles across town did seem like a lifetime away. My father came to help me take my things to his place, but our ride there was silent, and that silence would extend well beyond that trip. Not once did I find myself rocking in Bishop House, for I refused to find comfort there.

Shortly after graduation, I turned eighteen and was allowed a quarter of monies left me by Mama B. I had no idea that the money was coming; it had apparently been set aside since Papa Bishop's death. "Like she always said, funeral homes are a good business," Aunt Jess had said as we sat in the lawyer's office. This made my father finally see that his "support" wasn't something that I depended on. He never spoke of this trust, and he seemed to resent it.

On the evening of my graduation, I sat on the football field, looking into the stands trying to find his face among the crowd, having been assured that he wouldn't miss this day as he had missed so many others. When our tassels had been switched from left to the right and most caps soared into the night, hailing a new beginning, mine remained on my head, still clinging to the past. Yes, I had wanted to release my ties to the past and the present, yet doing so is a different thing altogether.

After the ceremony, Carol, Aunt Jess, Jason, and Jessica found me. Pictures were taken; I even smiled for them, hoping to hide the fact that my father's absence had cut to the marrow. I sent them on their way, assuring them that I had a full night ahead of me. The crowd slowly began to disperse. After the last flash had flashed and the final yearbook had been signed, I walked up to the stands and sat there, watching as the janitorial staff, distinct in their gray-blue overalls, took away the chairs and broke down the equipment. Everyone had moved on to preplanned celebrations, but I remained in the bleachers.

I was told that they were locking up and the lights would be turned off. I said that that was fine. I was well known by then at BH and so had acquired certain privileges that most couldn't. I was the black kid who didn't play sports but somehow still managed to command respect that extended beyond football and basketball seasons. The janitors, black like me, had taken me under their wings, so they bent the rules now at my request. After their wishes of good fortune for me and our handshakes good-bye, I was allowed to stay in the stadium.

The buzz of the field lights shutting down sounded like the ending of life. No band played, no crowd waved pom-poms and oversize number-one hands or chanted for a TD. No students drank Bacardi and Coke. No elementary-school kids ran around in their little BH T-shirts, dreaming of the day when they would play on the field rather than in the sidelines of imagination. There was no one there but me, sitting in my gown, my mortarboard dangling in my hand.

I climbed over the railing of the stands and walked across the track onto the football field. I ran around with the intensity of a leopard chasing a gazelle, swerving and braking. I savagely ran that field until I was completely out of breath. I fell to the ground on what would have been the fifty-yard line, where the bear mascot emblem would have been branded into the ground had the white battle lines of fall been in place. I felt the pampered grass under my hands and thought to myself that football fields got the same amount of attention as lawns—often more attention than those who play on them.

I began to pull up the grass. Not in a violent way, but blade by blade, like one might when lying next to a new lover on the bank of a stream. I let the wind take the blades where it would, but they just seemed to find their way back onto me.

I was reminded of the night that Paul and I had come out to this very stadium after his brother's funeral. We climbed the fence in our suits. The space and solitude consumed us; death is a solitary act, yet a crowded affair.

"What was he thinkin'?" asked Paul. "Why did he have to . . .?"

Of course, I had no answer, and I don't believe that he expected one. We sat there drinking Miller Lites and throwing the cans around as if they were the miniature footballs tossed to the fans at halftime.

"Have you ever thought about doin' it?" he asked, looking wherever and beyond.

"Yeah," I said. "Who hasn't? But thinking about it and doing it are two different things."

"But would you do it?"

"Nah. Too afraid."

"But if you did do it, how would you?"

"I don't know, drown, maybe," I said, recalling my bout with the Pacific.

"I hear that's the worst way to go."

"We're talking suicide, you dickhead, not a summer social. It's not supposed to be fun."

"Touché," said Paul. "Oh, man. I just don't know."

"Sometimes it's probably best not to know."

"You remember Whitney? She wasn't afraid."

Whitney Cooper Braun probably was afraid, but she did it anyway. Right at school. With a name like Whitney, I wondered how bad things could be. She was popular and "seemed to have it all together," her friends had said to the reporters. Just once, I wish people would say, "Yeah, I saw it coming." But of course, that would imply they were somewhat to blame, and no one wants that.

After Whitney's funeral, the first of what would be two that year for students at our school, the school board decided that we should all go through counseling. That seemed to do more harm than good, for it kept the topic fresh while evading the subject all together.

But Mrs. Brown, who taught biology and was my favorite teacher freshman year, took a different approach. She wasn't as traditional as most of the teachers who had taught generations of Bears, and she was often shunned for it. She knew how to relate to students on their level. When asked if a test could be postponed, she'd say, "Sure, sure, weather permitting. But according to my calculations, I don't think hell is freezing over anytime soon."

A few days after Whitney's funeral, we were all in third period, in our alphabetically ordered seats, yet one seat was glaringly vacant. Mrs. Brown knew our thoughts were still focused on Whitney and that we weren't digesting the lesson *of the day, so in the middle of her lecture, she placed her pointer on her desk and ushered everyone into the lab. I found myself thinking* Damn, a pop quiz, and I didn't do the homework. *At each work station, she laid a dissecting kit.*

"Is anyone else in here considering killing themselves?" she said as casually, as if she were speaking about the weather. "Come on. 'Fess up. If you are, just let me know now." All eyes were on her as she picked up a scalpel. "Come on, speak up. Because if you are, I just want you to know that if you're considering slicing your wrist, as that seems to be what you youngsters prefer, you have to cut down, not across. Because if you cut across, you'll only bleed a little and pass out, missing all the major arteries, then some responsible person is going to have to clean it up. No. If you feel you must do it, then do it right. You have to cut down! Not across! Down! Got it?!"

She put the scalpel back on the table in front of her and just stood there, but she wasn't at all still, for her entire being was trembling. No one knew what to say.

Silence.

Then she started to cry, and then to weep. I left my seat and walked up to her desk. I tentatively put my arms around her, and she found the strength to raise hers to my shoulders. We held each other and no one made a move or a wisecrack. Moments later, everyone in the room started crying, from Blake Jolibois, the only freshman on the varsity football team, to Lloyd Pennington, the class brain. And like two insane people, Mrs. Brown and I looked out on the room at everyone else crying, then looked at each other and started laughing.

Slash down, not across.

Paul and I had laughed at the memory of that biology class, rolling through the grass on the football field, trying to distance ourselves from his brother's death, but it was the memory that was humorous, not the situation.

✦   ✦   ✦

I finally left the stadium that night. Still draped in my graduation gown, I drove to Mama B's grave, the third tree next to the fence. I placed my mortarboard on the granite and then went home.

My father didn't go to the airport when I left for New York. Carol said he just couldn't stand to see me go. I told her that I knew how he felt.

Carol was in the attic when Jeremy and Jessica returned. The room was so spacious, almost three times as large as Jeremy's first Manhattan apartment in the East Village, where many, like him, paid exorbitant rents so they could hold the title of the privileged poor. After years of playing the role with ramen noodle precision and ruining his feet wearing the prerequisite heavy black boots, he moved to what many of his slumming cohorts considered the suburbs—the Upper West Side.

Jeremy climbed the stairs. A sob drew him there. Carol was sitting on a box, a photo album open across her lap and a hand-kerchief balled in her hand. When Jeremy said her name, she immediately stood, turning into Carol Bishop, Attorney-at-Law. She wiped her eyes.

"It's dusty up here," she said behind a false smile. "Oh, who am I fooling? I thought I could handle going through these things. I don't know why I'm up here."

"It's okay."

She looked at him as though she had been waiting for him to say those words to her for as long as she had known him. Her eyes could no longer contain the tears.

Jeremy walked over and embraced her. He held her tight, the way he imagined his father must have in those moments when the shell she wore cracked, revealing the soft insides. He said nothing. He just held her, then caught her when her sobs were too heavy for her own legs to support. He held her up, and though she wasn't his mother, he held her like he would have held his own. He had rather die than see his Mama B or Aunt

Jess cry; now, for some reason, Carol brought out that response in him.

When the sobs subsided and the tears stopped flowing, all that filled the space were two strangers drawn closer and the dust-covered boxes of memories that had served as impetus.

"I want to show you something."

Carol walked over to the nearest corner of the attic. She pulled back the dropcloth and started taking down other boxes. She brought one box to where he stood. She told him to sit, and he did. She began to open the box, then stopped.

"I know I'm not the one to tell you this," she said, her hand on his. The dampness of her handkerchief added to the moisture already coating his perspiring palm. She removed her hand then continued. "Your father loved you more than I can ever tell you. More than *he* could tell you." His heart began to hammer a staccato, and he sucked in air through his mouth as his jaw fell.

Jeremy stood up. He wasn't ready for another lecture on the goodwill of his father, not even from her. He had heard it from every woman who'd known him, each trying to rectify or justify his father's actions. Jeremy was tired of being told his father's story by others and wasn't even certain that if the man whom they so gallantly defended was still here that he would want to hear it from him. But had his father been around, the subject wouldn't have been mentioned, for it was because of his absence that Jeremy found himself listening to Carol reiterate familiar phrases.

"Please, Jeremy," she said, almost as if she was asking more for him than for herself. "Sit."

He walked back to the box and sat in front of it, but sideways. Maybe if he didn't face it directly he could avoid what she was saying, avoid seeing her, and avoid all that was happening.

"Perhaps it was wrong of me to say that," she said, suggesting a lawyerly fairness. "I'm only a witness, not the victim. I'm a step away. I'll never know what you've gone through, and I don't want to sound condescending. He was my husband and I loved him and will defend him to the . . ."

Her voice refused her as she thought of the word she had been about to say. She paused for a moment, and then even chuckled at the circumstances. She had delivered so many summations and knew how to use each word for its paramount effect, but here, sitting on a box in the attic, she was faltering.

"When I met your father, he just looked like any handsome man. I'd never been married, never really dated. I was fresh out of law school and had always believed I had to prove something to the world, for that was the only way the world would accept me. I didn't really know how to love—just how to study.

"On our first date, he told me that he had a son and that I needed to know that up front. He showed me your picture, a school photo, I believe it was. He was so proud."

She reached down for the box and Jeremy looked out of the corners of his eyes, still not wanting to face what lay inside.

"Your father didn't know that I knew about this box," she said, looking at the box as though it were alive. "I never told him that I did and never bothered him when he came up here on those nights when he couldn't sleep because he was thinking about you. He'd never admit that, but I knew you were on his mind. I would hear him pacing. Most people pace when they are thinking or nervous, but your father did it when he was hurting." Jeremy listened, and gradually his body began to shift. Carol pushed the box over to him with her foot and then got up.

Jeremy watched her every step and met her eyes when she looked back one last time before descending. He started rocking on the box on which he was sitting, and he stared at the one now in front of him. He was four flaps away from his father's concealed past. His eyes scanned the attic. He wanted something to preoccupy him, something to focus on other than that box. He wasn't ready to face its contents.

*   *   *

When Jason and Jeremy pulled up in front of Jess's they looked at the multitude of cars that surrounded the yard. Of all of them, one stood out. It was the pink Cadillac with a Mary Kay sticker on the rear windshield.

"Aunt Gladys," they both said.

Gladys was the middle child and lived in Dallas with her husband, Laron, who worked for American Airlines. For years, everyone just assumed he was a pilot, for that's how Gladys made it sound. But it turned out he was a maintenance man. She had kept that from the family; she was ashamed of him. "I married down, as a lot of women do," she said with a sigh when the truth was revealed.

Laron rarely accompanied her on visits to Elsewhere. She always made up some excuse as to why he couldn't be there. A Thanksgiving visit years ago was the reason for this; she never invited him back. Laron had been having a great time staying out of his wife's way, enjoying the fact that she had other distractions. When the meal was placed on the table in all its glory, Gladys became unstoppable. Her mouth was moving nonstop, with a bite—but not the kind associated with food. She went on about how wonderful her side of the family was, throwing her comments Laron's way, like one shaming a child. She kept going on and on about her "family money," so that it became embarrassing for everyone in the family.

Finally, Laron had had enough and he spat out, "Well, that's all fine and good, but you married to me now, and that means my family is your family, making you just a first cousin away from government cheese, so you can just can it with the how great your family is!"

This closed Gladys's mouth like a turkey sewn up after being stuffed. Everyone was relieved by his outburst and began to pass the plates of food around at a quicker pace. Seconds and thirds were had as Aunt Gladys remained tight-lipped, with no wishbone in sight.

*Even Mama B had once said, "I don't know where I went
wrong with that chile."*

"Don't make me have to go in there and deal with her alone,"
said Jeremy to Jason. "You might as well get it over with now, or
you're gonna have to deal with it tonight, in public."

The two of them walked toward the house, which they could
see through the window was filled with people laughing and slap-
ping knees without regard. When the two entered, the noise level
rose. Jess sat in her rocker, beaming like a proud mother, looking
at her two "young and handsome" nephews.

"Me? My goodness. Me, is that you?" Gladys called to Jeremy.
Of all his names, he hated that one most. But since he seldom
had to deal with his aunt, he never bothered to voice his
opposition.

"Hello, Aunt Gladys," he said. As she made her way toward
him, he pulled Jason in front of him. "And here is Jason. He was
just asking about you on the drive over."

Jason took the hug meant for Jeremy but gave him a look that
more than said retribution was forthcoming.

Their cousins were standing behind their mother. They were
in their early twenties, still dressed alike and holding steadfast to
the three childhood ponytails atop their heads.

"Tamara? Tameka? Can't y'all come say hello to your famous
cousin?" They didn't move. Waves, not hugs, were exchanged,
but that didn't sit well with Gladys. "We're family, and families
do what? They hug. Now go on over and give your cousin a
hug." Robotic hugs were exchanged.

"Yes indeed. Just two fine young men. I guess it's up to the
two of you to continue the family name. I really wanted boys,
but you get what you get, and I shole ain't havin' any more babies.
I suppose I could; women havin' babies much later nowaday. I
saw this segment on *60 Minutes* about a sixty-three-year-old
woman that gave birth. What do you think about that, Me?"

"An amazing piece of work," said Jeremy with a hint of sarcasm. "No, Aunt Gladys. I don't think the world would suffer if you didn't have any more children. How's Uncle Laron?"

Gladys's face stiffened and Tamara and Tameka looked at their mother.

"He's fine," she said tersely. "Too busy to come. I had to drive all the way myself. I don't trust the way these girls drive. I tell you, you send them off to college and they forget all the home trainin' I taught them. They think they grown."

"Well, they are, Gladys," said Jess.

"These here are *my* children, Jessica. I know you're an expert at rearin' other people's children, but these are mine and I'll decide if they grown or not. Thank you kindly."

Gladys had always talked down to Jess and for that matter, to everyone else who didn't seem worthy of her presence.

"Well, Me. I know you're goin' to take my picture before I leave, ain't you? I told everyone at Mary Kay that I was goin' to have my famous-photographer favorite nephew take a picture of me."

"I'm sure they will be powdered with disappointment," said Jeremy, still annoyed by her, but not wanting to cause a stir, which she was more than capable of starting. He added, "I didn't bring my camera with me."

"Tamara, Tameka. Make yourselves useful and get Mama's camera out of her purse."

"Gladys, it's gettin' late," said Jess. "The boy didn't come down here to work. He's got to get ready for the wake."

"Oh, hush, Jess. You're always tryin' to monopolize him, keepin' him away from everyone else. I see you got your picture up there on the wall. You don't mind, do you, Me? It's the least you can do. I don't mean to bring it up, but that was *my* daddy's money Ma Dear gave you. A piece of that would have rightly gone to me and mine, so I think you at least owe me a picture."

"Gladys!" screamed Jess, springing from the rocking chair as though pressure had launched her. "That money was Ma Dear's

and she gave it out as she pleased, and you sure weren't com-
plainin' when you cashed yours in."

All other conversations came to a standstill, but Tamara and
Tameka reappeared, Tamara holding a disposable camera. She
handed it to her mother, who passed it on to Jeremy. Jason looked
at his brother and started laughing. He didn't attempt to conceal
it. He laughed all the way down to the floor, then rolled on his
back at Jeremy's feet, like a dog longing to have its belly scratched.
Jeremy didn't try to explain the situation to Gladys as she posed
in front of Tamara and Tameka. He looked at the camera with
a smirk and placed it up to his eye.

"Say 'government cheese!' "

Flash.

chapter 16

∽ ∾

J eremy stood behind Johnson's Funeral Home. He wasn't ready to go in just yet. He had wanted to walk in with the others, but he told them he needed a moment, and no one questioned him or denied him that wish.

He squatted against the brick exterior, arms perched on his knees, looking as though he were posing for a picture that he would shoot. Wistfulness and ambivalence consumed him, fitting his body as perfectly as his tailor-made suit. The hum of the central-air unit was his only companion as he watched the cars pull into the parking lot, hoping that their headlights wouldn't stare at him too harshly.

After a last few seconds of mental preparation, he finally walked around to the front and found his way through the perpetual portals. The lobby was overflowing with people greeting one another with alacrity as though it were a convention, rather than a wake. Vague faces spoke equally vague words. Everything appeared to him as a blur, like movement caught by a long-exposed shutter. He could see lips moving, but all he could hear was the gospel Muzak gnawing at his ear. He shook hands, smiled, and hugged indiscriminately, like he might do at social functions in New York, but instead of making business transactions, here he must say all the proper things about loved ones.

"We're just so proud of you."

"What a fine tribute you are to your father."

"It's just been too long since your last visit."

"You remember me? I was in your homeroom in fifth grade."

Young women his age and the inversion thereof pointed, giggled, and gossiped like debutantes hoping to become Mrs. Jeremy Bishop and be swept from Elsewhere to New York City. He met some of their smiles unknowingly.

He turned and looked across the lobby, and he could see the casket from another wake, yet no one appeared to be entering or exiting the room. He wanted to walk through the crowd and sit with the stranger, show him some attention, to get the focus off himself. He wanted to know who this man was and why it was that no one was standing over his casket. Even when others commanded his attention with conversation, he still positioned himself so that he could gaze into the other room where the lone man lay unnoticed.

Jeremy had been in the funeral home for ten minutes, yet he hadn't been in the chapel. He knew the lobby was a safe haven where trivial chatter stole focus, but once he entered the chapel, there would be nothing left for him to do but walk up for that last viewing, the first in years. How he would have loved to just walk in and tuck himself into the back row, but that was far from appropriate. He had to be in the front with the family: eyes would certainly be on him. He tried to find a way out of following the proper protocol.

But the time had come. He excused himself from those currently in his company, unaware that he was stopping them in midsentence. They would forgive him anything, for he had lost his father, and that was what was making him distant—it had to be.

He crossed the threshold and stopped. The group standing near the casket seemed to part when he entered, presenting him an unobstructed view. This casket also held a stranger. He gave the mandatory greetings to those in the back pews. With each greeting, his insides expanded like an earthworm surfaced by a rainstorm. Now, he was grateful for the delay that the necessity of greetings provided on this journey, for he knew what it all came to—a last look.

Jess and Gladys, as well as other family and close friends, were on the front row to the left, while Carol, Jason, Jessica, and their family were to the right of the casket. He looked at Jason and Jessica sitting with their grandparents. He watched as people entered, passing him on their way up to look at the body, then pass the handshakes to the receiving line of mourners.

Now it was his turn.

His shoes seemed to bury themselves in the sky-gray carpet. His steps were firm and deliberate.

A step.

The nose.

A step.

The forehead.

Step, the cheek. Step, look at one of the wreaths, look at anything. Step, the face. Step, smile at Jess. Step, feet together.

Stop.

There he was, in front of his father.

Time didn't exist as Jeremy stood at the casket. He felt a hand on his shoulder, but he didn't turn around.

He just stared.

He felt another hand on his arm.

Still, he stared.

He continued to stare as the body that owned the hands backed him away from the coffin and to the corner seat that had been vacated for him on the pew. His stare was so intent that he didn't feel himself moving back, didn't feel that he was sitting. He didn't feel the other hands that were rubbing his neck and shoulders and squeezing his biceps. He didn't feel the slaps of wind coming from the fan that one of the hands was waving at him. He felt many things, but none of those. He didn't see the circle of people now gathered around him. He just stared, the image burned into his mind; though suits of blue and black and gray blocked his field of vision, he stared still. He didn't feel the handkerchief being placed in one of his hands or the hand being placed in the other. He didn't hear the sobs of others that his stare evoked, nor did he hear the consolations of "It's awright." He didn't hear or see anything. He just stared. He stared until his eyes were completely dry. It was then that he finally blinked, and a tear took the focus from his stare.

"Have you ever tried walking through a mirror?"

That's what he had asked me at Christmas three years ago. I'd come home after almost five years. I had noticed the family Christmas card—my name was on it, yet not my face. I knew they weren't to blame for that.

The tree was up. It and the fireplace provided the only light. I'd been out, still avoiding him, even then. I had arrived and immediately after dropping my luggage in my room, headed out to find friends or go to Jess's, anything but be there.

"Have you ever tried walking through a mirror?" he asked, as I walked by. The blinking lights, mixed with the glow of the fireplace, illuminated yet made a kaleidoscope face.

"Excuse me?" I said, still wanting to continue on to my room, wanting him to say what it was he had to say so I could dismiss it with irreverence and be on my way. He had truly been making an attempt to reach out on this trip; I had come to visit at the request of Aunt Jess. I was older, I thought, and would make an effort, but old habits are hard to break, and the heart can't easily forget when pride is the aorta that feeds it.

"Carol and I went to a Christmas party tonight. There was a huge space with a door frame around it. I just assumed it was an entryway that led to another room. From where I was positioned, I could see people in the space. I wanted to see who they were, see what I was missing. I walked over and I could see myself in it, but I still tried walking through, thinking that there must be something more there. It never registered that I was walking into myself. There had to be more. But

then my nose was pressed against my nose, but not my nose, the mirror's."

It was as though he didn't know who I was while he was saying this. He was speaking out loud to whoever, but I was the only person present. He still had on his suit and his tie was loosened, collar unbuttoned, cuffs hanging. He sat on the sofa, looking at the fire as though it contained information.

"I looked around to see if anyone had noticed," he said. A quick exhalation pushed his head back, then forward again, an umph *without the sound is what the motion was. "But no one had, so I started grinning, like a kid who had gotten away with something and could finally be amused by it. But for the rest of the evening, I watched people doing the exact same thing—all trying to walk through the mirror, to see what was there. But there was nothing there, just a reflection of everything else."*

He turned away from the fire that imbued the room with warmth, then he looked at me as though I should understand his comments, but I didn't. I wanted him to say that he loved me or that he had made some mistakes. I wanted him to say everything that everyone else always articulated for him, but I wanted it all in his own words. I'd always wanted something from him.

Go. Come. Say.

Yet when he obliged, it was never enough, and I'd want something else.

"Have you ever tried walking through a mirror?"

"No," I said, removing the sting from my normally wasp-ish tongue.

"You should try it," he said. "It's very humbling."

He rose from the sofa and walked over to me. He raised his arm and I flinched a bit as he put his hand on my head. He looked at—almost through—me. We stood eye to eye, and I could see the specks of color from the Christmas tree lights twinkling there. I had been hypnotized by his words even though I didn't understand their meaning.

"You need some gas money?" he asked, stroking my head as if I were a teenager.

"No. I'm fine, thanks."

"One day, you'll have to say yes to me." He took his eyes from mine. Pulling the tie from around his neck, he turned from me.

" 'Night," I said, watching him walk through the door to the hallway. I stood transfixed in that spot as though he hadn't left the room. I looked at the lights on the tree, blinking in reverse time to the light within the black angel atop it, stealing focus from it. The small reflective ornaments captured the room in miniature. I walked to the tree and looked at a silver spherical ornament, seeing my face within its shine. I moved in close until my nose grazed it and my eyes crossed. I tried to catch it as it fell to the floor, shattering to expose its hollow insides.

Words were said, but not many. Jeremy looked over to find Jason, but he wasn't sitting next to Carol. He got up and crossed over to her anyway.

"Are you all right?" she asked, rubbing her fingertips over his hand.

"Yeah. I guess it just hit me. I'm fine, though. Where's Jason?"

"I think he went to get some air. It was getting to him, too."

Jeremy went to find him, softening enough on the way to assure the curious that he was fine and didn't need to be accompanied.

When he made it through the chapel doors, he looked around the lobby but didn't see his younger brother. He walked outside among the plethora of cars that had now begun to line the streets. He didn't call out Jason's name; he just walked around the parking lot to the back of the funeral home, where he found him in the very spot he himself had occupied earlier.

"Wassup?" said Jeremy, walking over to Jason, who tried to laugh while wiping his eyes. "Something in your eye?"

"Yeah."

"Yeah. That happens to me, too."

"It's a fuckin' zoo in there. Everyone's just actin' like he ain't dead."

"When you've been to as many of these things as I've been to, you begin to realize that funerals are for the living, not the dead."

"Whateva. I don't even know half of the people in there."

"I don't think we can expect to know everything and everyone. But they evidently knew him and . . . well, we all deal with things in our own way."

"I still just had to break. I wanted to stay there, for Mom, but I couldn't take it."

"I understand," said Jeremy, and he did. They went unnoticed by the group of people walking to their car. One of the women in the group was laughing, which seemed misplaced in the moment, for it was the only sound that carried to them.

"Listen," said Jeremy, "I'm curious. Did Dad ever ask you if you'd ever walked through a mirror?"

"What?"

"Walked through a mirror. He asked me that once. If I'd ever walked through a mirror."

"Nah. I don't think so."

"Did he ever tell you he loved you?"

"Nah, but I knew he did. I mean, you know. Some things you don't have to say, you just know."

"Right."

"You know, that's the first time I've ever heard you call him that."

"Call him what?"

"Dad."

The word hung in the night between them like the glow of headlights through mist, and Jeremy slid down the wall next to his brother. He put his arm around Jason's shoulders and held tight, searching for direction.

*   *   *

"Are my eyeth puffy?" asked Jason, taking on a lisp.

"You look faboo, dahling," said Jeremy licking his thumb, blending the tear stains—in the same way that his Mama B had done for him on so many occasions—on Jason's face, turning another page in their relationship. "Simply devune."

They walked back toward the lobby. Some of the cars had disappeared as people were beginning to disperse. They walked in and a man came over to Jeremy. He looked as though he had been crying, his eyes burgundy. He took Jeremy by his arms. It startled Jeremy, and he took a defensive step back, but the man held firm. He then hugged Jeremy. Jeremy's face scraped against the coarse fabric, almost burning his skin. The smell of mothballs mixed with liquor traveled to his nostrils, but didn't fog his mind. The face began to look familiar. He remembered the face, but from where? Then it came to him. It was the man from Mama B's funeral.

"It's all right," said Jeremy. "It's all right."

"That was my boy," said the man. Jeremy's body tensed. All eyes were on them. Jeremy tried to soothe the man, calm him. "I loved her, and they took the only thing that I had. They took him away."

Jess and Gladys appeared at the door of the chapel, and their steps freeze-framed as though a wall had been constructed between it and the lobby. They were deterred only momentarily; when they began to walk again, it was with the tenacity of a bedroom mosquito. They both took an arm of the man and pulled him away from Jeremy. Jess told Jeremy to go outside and wait at the car. He stood there, feet leaden. Jess repeated herself, and he and Jason started out. As his hand reached to push the door open, the man broke from Glady's grip and reached toward Jeremy, shouting "He's all that's left. That's my gran'baby there. Mine!"

His outburst took Jeremy back to the streets of New York and the illusions and disillusions that filled them. This man had that same distant, beyond-the-horizon look of those who didn't know where they were or even who they were and for whom words to themselves were their only possessions.

Jeremy walked back toward the man, as though confronting a thief before he could attack. "My grandfather is dead," he said, sickened by the thought of a scene, as the spectators looked on with disbelief. Jeremy tried to smile to maintain some sense of sanity, but he could feel his temper heating, mercury rising.

"I'm ya granddaddy," the man insisted, drunken tears pouring down his face like a leaky backwater still. Jeremy took alcohol to be the cause for this outburst. Jess, along with some men, ushered the man away, and Jeremy again reached for the door. But the man didn't go quietly. "They can make you go away if they want to," the man said. Jeremy turned around again. Then he looked at Jess, and her face held more words than she had ever wanted to say, more than she had already said.

chapter 17

◦∽ ∽◦

The house was finally quiet, if not at peace. Everyone had gone home full, eyes bigger than their stomachs, their departures choreographed by convention.

Jeremy went to the front room. He pulled his shirt out of his suit pants, examining the point that separated wrinkled from pressed; a fine line it was. When he had stripped to his underwear, he looked at the floor. He felt like the pile of clothes huddled there. He let out a soft moan as he lay back on the bed, pleased to be off his feet for a moment. No one had mentioned the scene with the man, as if it hadn't happened—story of his life. The house had been too full for him to find a moment to properly address it with anyone. He remembered asking himself just yesterday about the man he had encountered at Mama B's funeral.

He jumped from the bed, looking through the pile of clothes for his jeans. He slipped them on, and as he started to go in to speak with Jess, a knock beat him to the bedroom door.

"Come in."

"J?" she said, opening the door, yet not moving to enter. "You awright in here?"

"Yeah. How are you holding up?"

"Not too good," she said, leaning on the doorjamb for support. "I need to talk to you about somethin'."

Jeremy looked at her and slid, ever so slowly, back down toward the bed, as though every inch of the journey down and every muscle had to be explored. Her tone didn't hold its usual comfort, assuring him only that this wasn't to be an ordinary conversation and that he should give her his undivided attention. He patted the bed, motioning for her to sit next to him, but she waved him off.

"I think I better stand." A breath raised her breasts. "Lord, help me." Another breath and yet another. "That man at the wake . . ."

"I was actually just coming in to—"

"J, please," she said, her hands coming up to a prayer in front of her mouth. "Please, just let me say this while I'm still able."

Jess closed her eyes. Her hands went from her mouth to her temples. She rubbed in a circular motion and a vein forked like lightning across her forehead. She seemed to see the words and images flashing before her, squinting to find focus. When her eyes reopened, the words came out.

"That man *is* your granddaddy . . . not really ya granddaddy. . . . What I mean to say is, you see, he is ya daddy's daddy."

Jeremy didn't speak, but he was certain that she could see that his heart had stopped beating.

"I never thought I would have to tell you about this. None of us did. Ya father didn't even find out until a few years ago. That's when I asked you to come down for Christmas. He had taken the news hard. I thought your bein' down here would help him. But against my better judgment, I think I should tell you now."

Jess was still leaning in the door. Her voice remained tenable; she told the story as though testifying in court. She looked straight ahead, never directly at Jeremy.

"My daddy had been workin' so much, tryin' to get the funeral home up and goin'. Ma Dear met this man, though it was never known where, and she never told. I'd seen him, though. Ma Dear had denied that the child was the man's, but when he was born, it was pretty clear whose son he wasn't, and the denial became more of a way to keep the family together rather than defendin' what happened. Your daddy was given the Bishop name, but he wasn't a Bishop, and he was a constant reminder to Papa of what happened.

"He had to be a father to another man's son and never say a word about it. It was one thing to have his wife be with another man, but to have people know about it would've been worse. Divorce back then wasn't as common as it is nowaday. He wanted to make certain that his good name was protected. That's why he was never much of a daddy to ya daddy."

"And this man just let it happen? I mean . . . how did he take it?" asked Jeremy.

"I can't rightly say, J. It wasn't my place to ask. We all tried to stay out of the way of it. But it was believed that Papa paid the man to disappear, and when Papa said do somethin', you did it. This hurt Ma Dear 'cause she had already denied the man his child, and she and Papa stopped talkin' all together. That's when the rental house was built. Papa lived there until he died and Ma Dear never set foot in it. She was forced to sit over here and watch the construction. Every wall that went up moved them farther apart."

Jess had basically taken on the responsibility of helping raise the child who never understood why his father never looked at him and why when he spoke, no comfort was ever found in the words he chose.

Mama B had spoiled her only son, hoping to make up for what he never would have. The man was never mentioned. In time, life went on, and new secrets encircled old ones like the rings inside trees and the bark that covered them.

The only memory Mama B had of the man was the denim patch that sat with her in the rocking chair.

Jess finally walked toward him and sat on the bed, but she still didn't face him. The story had drained her of any energy that remained in her body. Even crying at the telling of it would have been too taxing. She could hardly believe that after all those years, the words had finally come out of her mouth, but because they had, an aura of peace prevailed.

Jeremy thought of his Mama B, his father, and his mother. Thoughts of all of them presented themselves in kaleidoscope, and he felt that he was doing the impossible, walking through the mirror to the other side, for once seeing more than just his own reflection.

"I always told you that your father loved you, and believe you me, he did. The best he knew how. We always made excuses for him 'cause we knew things he didn't know. Ma Dear so loved you. You were her chance to say she was sorry for things she couldn't rightly apologize for. This house has always held too many sorries never said.

"Your father never learned how to be a father, but when your mama was carryin' you, you could see that he wanted to be the best daddy in the world, be everything Papa wasn't to him. Then, well, then when your mama died, it took away a piece of him that he never got back. You can hate him if you have to, but now you know the truth. No more secrets. It's all out, so there's no reason for any of us to keep goin' through life with one eye closed."

Jeremy wanted to be upset with someone, everyone, yet as he culled his mind so he could pinpoint blame, he couldn't. All of a sudden, he found himself older than he was and younger than he'd ever again be.

Jess was in the rocking chair when he finally emerged from his room. She looked up at him. He smiled to reassure his aunt, but it wasn't a smile associated with joy—more a smile that said all it needed to say. He sat on her bed like he had done numerous times when Mama B had sat in that very chair, near the chifforobe.

"Do you know where my mother is buried?"

"No, J. I can't say as I do."

"Every time I visit Mama B, I look around, hoping I'll see a tombstone with her name on it, but I've never found it."

"He wanted to keep her to himself, and we gave him that. He

did call your mama's mama to tell her, and when she said she didn't care, your father broke down like a baby."

"Did she ever call or try to see me?"

Jess didn't answer, refusing to give him another no.

"No Christmas card or birthday . . . ?" Jeremy's voice trailed off, for he knew the answer.

Jeremy had retreated to the comfort of the yard, the lightning bugs' silent serenade entertaining him. He needed the fresh air. The house that seemed so large to him as a child now seemed small and puzzling. As he enjoyed the breeze, Paul drove up, parking his car behind the Oldsmobile. He walked over to Jeremy and sat down next to him. The ride of the rocking lawn bench stopped only for a moment before they both pushed it back into rhythm. They just rocked, back and forth, speaking periodically as thoughts came.

"For a minute there," said Paul, putting his hand on Jeremy's shoulder as he had done earlier that evening, "I thought you were gonna nosedive into the casket."

"I thought about it."

"I'm glad you didn't. Hell, if you do everythin' at the wake, what would you have to look forward to at the funeral?"

Their laughter bounced off the night.

"You ever notice how you always use the word *hell* and I always use the word *God* when we can't find the right transition?" asked Jeremy.

"No. I've never thought about it. Daniel always said *hell* so I guess I picked it up. Product of my environment."

"Maybe. Whenever I go to or watch one of those award shows, the black winners always thank God first. Always. But I can't recall, not even once, a white winner thanking God."

"Maybe black people have a different belief system, different priorities than white people."

"And where do you fit in?"

"Hell," said Paul, "I don't know."

"Me neither."

"It's a beautiful night," said Jeremy, passing the Philly Blunt–sized joint back to Paul. "I shot this physicist, for *Time* magazine, I believe it was, and he said that Darwin believed that without black holes, there would be no life. The world couldn't even exist without these mysteries filling the sky." Jeremy stared at the sky, amazed, as though he were viewing it for the first time. "So many stars. Look at them."

"Like they don't have stars in Manhattan?"

"Yeah, they do. I just never look up. It's hard to see beyond the glare."

"I guess it is easy to take them for granted. All I ever look for is the Big Dipper and the L'il Dipper, and when I find them, you'd think I'd won the fuckin' Nobel Prize. I just point them out like a new discovery."

"You know, there's this place in Illinois that, for forty-five dollars, shipping and handling, you can name a star," said Jeremy.

"Getthafuckout."

"No, I'm serious. I read it in the *Farmer's Almanac.*"

"You read the *Farmer's Almanac?*"

"Yeah, I keep one in my bathroom. It's full of information."

"You're full of somethin', all right."

"Whatever," said Jeremy, adding, "You think the other stars get jealous of the moon?"

"I doubt it. They have no feelings. And besides, that's the great thing about stars. They have names, but you don't have to know them to appreciate them. They're just there to be looked at, from a distance. Kind of like you."

"What do you mean by that?"

"You were always the star. Everybody knew it and appreciated it. Everybody respected you, the teachers, the preps, the nerds, the freaks in the smokin' section, the band fags, the choir fags, everybody. You treated everybody the same. But none of them ever thought they could get close to you. The only way they could get next to you was by castin' their vote for you, so they would. It was like you were just too far away. Look, but don't touch."

"You're stoned."

"Of course I'm stoned. But I'm tryin' to be serious," said Paul, a bit irritated. "Why do you always do that? That's exactly whatthafuck I'm talkin' about. You always alienatin' people. They say somethin' and you give some flip comment that makes them feel like what they're sayin' doesn't matter. Like they're stupid and should just be blown off."

"Paul, that's ridiculous."

Paul turned away from Jeremy and shook his head.

"You know, tonight when I was standing up there, it was the first time I'd ever really looked at him. I mean truly looked at him. I'd always looked around him. I never wanted to look at him. I didn't want to see any resemblance, you know?"

"Yeah, I know."

"It was like he wasn't my father at all. It was like he was another man. But tonight, for the first time, I didn't look away. I didn't want to avoid him. I just wanted to see him for what he was—just a man."

"That's usually how it goes. I hated that Samantha and Daniel had a closed-casket service for Pat. I kept lookin' up there at that fuckin' photo of him smilin', and I just wanted to run up and rip it off. Even though I was mad as hell at him, I wanted to see him again, 'cuz I knew it was gonna be the last time.

"I hated them for months for that, but it was different for them than it was for me. I wanted to see my brother, but what they would see wasn't just their son. I know every time they look at me, I'm just a reminder too, you know?"

"I know."

❉   ❉   ❉

"In a few days, you're gonna be a pop."

"Yeah, ain't that a laugh?"

"You're gonna make a great father."

"I don't know, you know?"

"Don't worry. I know."

"When you headin' back?"

"Day after tomorrow."

"Why can't you just say Sunday? You always get so dramatic when you stoned. 'Day after tomorrow.'"

"What are you talking about? I only got one good hit—you were over there hogging it, ya punk."

"Yo—a punk is a piece of wood."

"You know almost every picture I've seen of yours has been black and white? You really oughta put some color in your life."

"I just prefer black and white. It transcends time and leaves something to the imagination. I tried shooting color when I got to New York, but when you develop color it has to be in complete darkness. No light whatsoever. You don't witness the magic. When you're working with black and white, there's what's called a safe light that provides a glow, and you dip the paper in the baths yourself. It's more hands-on. It allows you to see the image as it appears."

"So what you sayin' is you afraid of the dark?"

"Maybe. I never thought of it that way."

"Yeah. It's nice out."

"Yep. Thanks for coming over. I appreciate it."

"Ain't no thang."

"You're crazy."

"Yeah?"

"Yeah."

"I love you, man."

"It's the pot talking."

"No, it isn't. I mean it. I love you."

"I love you, too."

"You're the first man I've ever said that to."

"You're the first man who's ever said it to me."

Paul got up and walked to his car. The engine turned over and the headlights lit up the yard as he backed out and drove away.

Jeremy felt movement on the back of his neck. He put his hand back there and discovered a small bug. Without a thought, he squeezed it between his forefinger and thumb. He pulled it around to examine it. The body was smashed, but its light continued to shine.

It had been a cold winter in New York. "The coldest in several years," the weatherman had said on numerous occasions. But I couldn't rightly say, for it was my first winter in the city. But yes, it was cold.

"I ran a bath for you," he said.

I stepped into the tub. The water was hotter than my hand had forecast, and it scorched my skin as I eased into the tub. But the temperature didn't matter. I wanted it as hot as I could stand it. You see, it was as though I was just going through the motions, for that was all I had.

The water was clear. No bubbles or oils. Just water and me.

Music was playing, as it had been all night. Sweet music. Music for lovers, embraced in wordless passion. It sounded so sweet, each note its own, yet joined by another and another. The scent of something raspberry filled the tiny yet homey apartment. The music played on.

He politely knocked on the bathroom door, a cool yet warm smile on his face. He was carrying a tray with a pot of tea and a cup on it.

"Raspberry Royal," he said. He put it on the lid of the toilet. "It's still brewing. Give it a couple more minutes."

I couldn't speak. It all seemed so strangely commonplace, as if this was the way we had done it each morning for years.

He ran his hands through the water, never touching me, but just close enough that I could feel the pressure of the water caressing my body. I closed my eyes. Scooping up water with his hands, he poured it over my head. Still hot, but comfortable, the water ran down my back. The back that not so long ago had been held in his arms, tightly.

"See, that wasn't so bad, now was it?" he said, kissing the top of my head and getting up to leave. He spoke just above a whisper, but I could hear every word. I stared at the tray of tea, amazed. What did I do to deserve this? Why was he doing this?

Shortly after, I got out of the tub, steam rising off my body. I quickly dried myself off, attempting to avoid a chill. I went into the bedroom. He sat reading the New Yorker and sipping his tea. My clothes were still in a pile on the floor, where I had placed them the night before: coat, shoes, socks, shirt, thermal, T-shirt, pants, thermal. I got dressed: thermal, pants, T-shirt, thermal, shirt, socks, shoes, coat.

It was cold.

"I'm leaving now," I said.

"Oh. Have a good day," he said. "Last night was great."

That was it.

On my way out, I glanced into the bathroom to make sure I wasn't leaving anything I needed. The water had drained and only a ring remained. I left it there for him to see. On the toilet, the tea was still brewing, captured breaths dancing their way through the spout. As I stood at the door, it all looked like a picture; black-and-white silver gelatin print.

When I got out of the building, the hawk of winter grabbed me, and I could feel my skin drying beneath the layers. I hailed a cab, told the driver where I was going. He was in a jovial mood, wanted to talk, but I didn't much feel like saying anything. I couldn't. I just sat there, staring at the dawn, watching the buildings doing their best to discourage the eastern sun, only to be defeated. It all so reminded me of the *night before, and the cab ride made me long for home, for something familiar.*

"Second and St. Marks?" said the cab driver, bursting a bubble of thought.

"Oh, yeah. Thanks," I said, not realizing we'd reached my destination. I slipped the money through the slot and got out.

"Have a good day," he said, and the cab rolled away.

The driver didn't have to venture far—someone on the next corner was waiting to be picked up, and off they went.

When I got into my fourth-floor walk-up apartment, I stayed in the bath for most of the day. When the water became cool, I'd just add more hot water. The mirror steamed over, making it impossible for me to see myself; there was only an impression of me.

I never told anyone. What would I have said? I was almost nineteen. An adult. I had left that art opening with him of my own free will. Who would have cared? And maybe it wasn't really rape. I had made the choice to be there. No one would believe I said no. It just doesn't work that way.

As with so many experiences in life, it was left unsaid, and I remained wondering and wandering, with only my rocker to turn to, and no love lost.

Usually, every Sunday night before he fell asleep, Jeremy would mumble to himself, "Another week." But it wasn't Sunday, and the days had seemed to run together like the threads and pieces of colored cloth and denim.

His body ached. He felt abused and bruised, for the day had proved arduous as secrets surfaced. He ran the bathwater but didn't notice the things that decorated the room this time.

Jess hadn't said anything to him, nor he to her, when he came in from the yard. She had her sponge rollers in her hair and a black dress hung on the door. The plastic that covered the dress captured all the light in the house, making its contents appear alive.

He took his black suit out of the closet and hung it on the back of the bathroom door. He sat on the fuzzy top of the toilet seat, letting the steam from the bath fill the room. Sweat began to glaze his body. He took his finger and ran it across his forehead, then along his stomach like a windshield wiper, clearing a path that would soon be filled again. He looked at the suit on the door, knowing that in the morning he would fill it for the funeral.

He turned off the water and sat on the floor by the tub, his back against the cool metal and his arm resting on its shoulder. The candlewicks blossomed; the room was lifeless except for the dancing shadows they cast. The crack under the door glowed, but that was all that existed of the outside world for him. He just sat on the floor. He wasn't feeling sorry for himself, as he had done so many times before. Now, he felt sorry for everyone else. Bodies had finally been cut open, revealing the cancer.

The steam, too heavy to stay on the mirror on the wall, began to condense into a map. He looked up at it, watching the trails, and none of them made a straight line. He didn't see the drops as tears inching their way down. No, they again made him think of the stretch marks that covered his ass, growth in every shading, an indelible mark to show that the skin that covered his body at one time wasn't large enough, like a suit too small, so it stretched to accommodate, taking and making another form. He'd felt the tightness of a suit before, but he couldn't remember ever feeling his skin stretch. Now, though, he feared he could.

Their voices, all the ones he'd heard and the names they had given him, ricocheted off the bathroom walls and eased their way, on the back of the steam, through the crevice at the bottom of the bathroom door, away from the heat and the salty sweat that attempted to tender his soul.

He rolled onto his side and placed his head on his arm, still trusting the shoulder of the tub. He watched the drop of water on the faucet beat like a heart getting larger and larger, until it fell down to make the circle that would skim across the rest of the fallen drops. Even when the water came out in a hurry, with great force and pressure, he knew it was all nothing more than individual drops joined together.

He ran his fingers across the water, puncturing the surface, which made his fingertips appear larger than they were. Then he slapped his palm on the surface of the water, and it felt like needles pricking his skin. His hand made its way down deep and pulled out the stopper. When the water began to escape, he convinced himself that it was the ocean and the waves were more powerful than he; though he could appreciate it, he didn't want to go beyond the shore. He just sat by its side and listened until gravity had made its last sound and the water was gone. No ring was left, no sign of anything.

Though he wanted to stretch out on the floor all night, he got up. He washed his face and took the bath towel and wiped it dry. He turned to the mirror and wiped it as well, but the moisture remained, denying him an unconditional look. His suit still hung on the back of the door. He looked down at the spot on the floor below the suit, as though wanting to see the wrinkles huddled there, but there was no trace of them.

He opened the door. The cool rushed in like a seaside breeze in October. The wrinkles about his brow too were gone. He had bathed himself in his own sweat, leaving whatever it was that remained in him on the floor of the bathroom.

chapter 18

∽ ∾

"**Y**ou're up early," said Jess, pulling the cigarette away from her mouth. "How'd you rest?"

"Like a log. I was exhausted. Can I bum one of those?"

"I'm really tryin' to quit."

"Me, too."

She handed him her leather cigarette case and the lighter. He pressed opened the clasp and opened it, exposing the Viceroy 100's. He pulled one out and lit it.

"You know, Ma Dear smoked this brand," she said.

"I remember. I'm sure I racked up thousands of miles walkin' to Scalia's to pick them up. 'When they get to be seventy-five cents, I'm quittin'.' "

"Now they're a dollar seventy-five."

"Two-fifty in New York."

"Cost of livin' and dyin' is risin'."

"You can say that again," said Jeremy as they sat there, comfortable with the old levity. Jess looked at him almost as though she was amazed by what she saw—as if he was a phenomenon she couldn't quite understand, yet wished she could.

"You were the thread that kept this family together."

Jeremy's sarcasm-lined smirk was followed by a grunt, and he shook his head as he said, "I don't know about that."

"Well, you oughta know it."

"It seems like I was nothing more than a reminder of things people would rather have forgotten."

"Yes, you were that, too," she said in a mother's tone, acknowledging his recognition of the truth. "But what you reminded us of was the love that tied us together. Each house holds someone or something that makes a difference. You were that person. But sorting the wheat from the chaff takes time. Remember that.

"Today, we're buryin' ya daddy. Don't let ya feelings for him be buried, too. I can tell you he loved you until I'm blue in the face, but I can't make you believe it."

"Actions speak louder than words, Aunt Jess."

"So does silence, J. So does silence."

Gladys came in from the bedroom, Tamara and Tameka behind her like backup singers told not to shine.

"Mornin', Gladys," said Jess. "How'd you rest?"

"That mattress is a killer," Gladys said, tightening the belt around her pink bathrobe, which was accompanied by matching slippers.

"Good morning, Aunt Gladys."

"Says who?"

"Well, you made it through the night; that oughta count for something."

"I suppose," she said, waving her hand in front of her face to deflect the smoke. "Now, Jeremy, I know you're probably all out of sorts about the news of your precious Mama B, but I'm glad it's out in the open now. There's knowin' of and bein' of. I told them they should've told you years ago, but you can't tell these people nothin'."

"I appreciate your concern, Aunt Gladys, but I'll survive."

"I just thought that when you got the first piece of your money, you should have known where it came from. That was family money, you know. Papa worked hard for that money. Just you remember that."

"Gladys," said Jess in a whisper of contempt.

"Now, Jess, I've said my piece and that's all I wanted to say.

That's the problem around here—nobody wants to tell it like it is."

Gladys went into the bathroom, and Tamara and Tameka stood there as though they were waiting for their next command.

They drove to his father's house, where the limousine was to pick them up. Jess and Jeremy rode in Jess's Town Car and Gladys and her girls followed in her pink Cadillac. "You can ride with us; there's more than enough room," Jess had said, to which Gladys had replied, "I'll take my car. It's free advertising for the company."

"Your Aunt Gladys is a bitter woman," Jess had said when they got into the car. "That's a horrible thing to have to say about your own blood, and I love her, in my own way, but I'll never understand her. She has nothin' good to say about anythin'. That's always been her way. Don't let it get to you."

"I won't."

When they pulled into the driveway at Bishop House, Jason and Jessica were outside. They looked older than their sixteen years.

"Look at you, Jessica," said Gladys, pinching her cheek. "Just as pretty as you wanna be. I truly think your daddy should have named you after me."

"Well, you should be sure to mention that to him today, Aunt Gladys," said Jessica. "I'm sure he'd love to hear it."

After a beat, everyone began to laugh, even Tamara and Tameka. Gladys looked at them but that didn't subdue their laughter. They didn't even cup their hands over their mouths to conceal it, for their laughs were too wide. Gladys left them all outside with their merriment and went inside. She'd chosen not to wear black like the rest. She wore her favorite Mary Kay color, pink.

"Now, Jessica, you know you shouldn't have said that," said Jess, in a reprimanding tone, yet as she passed her going toward the house, she whispered, "But I'm glad you did."

"Wassup wit' y'all moms?" asked Jason. "Who put the crabgrass up her ass?"

"She did," said Tamara, and they all laughed again. It was the first time either had ever said anything about her mother to others. Perhaps even they had had enough. They seemed more distant from the family than Jeremy. They were merely family by association.

"I don't know how Uncle Laron puts up with her," said Jessica.

"He doesn't. They've been divorced for years," said Tameka. "She forces us to come in these little outfits to assure 'the family' that everything's fine. These clothes alone say there's a problem bigger than their being divorced."

Jeremy hugged Carol's mother and father. He had met them several times. He watched them with Jason and Jessica, and no envy came this time. He was pleased for them. They were there for their daughter and grandchildren, but didn't intrude and remained off to the side. Together, they all appeared to have been the family he had so longed for, both sides represented, and though he felt that he wasn't included in it, he knew that he was.

His thoughts climbed the stairs to the attic, and his body soon followed suit. He opened the door and turned on the light. The box wasn't where he had left it. He went over to the corner and picked up the dropcloth, but the box wasn't there either. He scanned the room, but there was no sign of it. All the boxes were gone. He went back downstairs and to his room. It wasn't there. Love and proof of it had again disappeared.

When the limousine arrived, Gladys inspected the car and got in first, motioning for Tamara and Tameka to get in. Jess told Carol's parents she felt they should ride with their daughter, so they got in as well. Gladys turned up her nose and moved her dress over as though she didn't want them to touch her.

"Jeremy, are you sure you don't want to ride with us?" asked Carol, standing at the door of the limousine.

"No, you and Jessica go on. Jason and I will drive behind."

"Okay, then." Jess got in, then Carol, the door closing behind them. The tinted windows took them out of view.

"You wanna drive?" asked Jason.

"Nah, you drive."

Jeremy and Jason got in the silver Jaguar, a car that hadn't left the driveway since their father died. Walter, the man who did the handiwork around the house, had washed and waxed the car to resale condition. It was spotless, just like their father would have wanted it. Jason turned the key in the ignition, but the engine didn't turn over. They looked at each other, then Jason tried again. The car started and they backed it out of the driveway to find themselves behind the limousine. They both had on their sunglasses.

"You know, he never let me drive this car," said Jason. "This is the first time. I mean, he'd let me back it out to let one of the other cars out, but this is the first time I've taken it on the road. He always said, 'As long as I'm alive, you'll never get behind the wheel of this car.' Kinda ironic, huh?"

"It tends to be," said Jeremy as he looked straight ahead to the limousine in front of them.

The limousine driver stopped in front of Macedonia Baptist Church. Jeremy and Jason found a parking spot down the street. Jeremy started to open the door, but Jason stopped him.

"Wait."

"What's wrong?"

"I was just, uh . . . forget it."

"No, what's wrong?"

"Do you think of me as your brother?" Jeremy hadn't expected this. Not today. Not now. "I just wanted to know before we go in there. He's our dad—both of ours—but do you consider me your brother?"

"Jason, I . . . don't know."

"That's cool." Jason went back into his aloof demeanor. They sat there for a moment in the luxury of their father's car. "I just know that I think of you as a brother. I mean, Jessica and I are tight. We've been together since we were born. But as much as I hated having to hear about you, I was still glad that you were my brother and I wished that I could know you better."

"So you're writing for Hallmark on the side now?" asked Jeremy. Jason took a playful jab at him. They started wrestling around the front seat of the car when the funeral director knocked on the window.

"Sorry to break up your fun, but the *funeral* is about to start."

"Oh, yeah. Thanks," said Jason.

"Cool. Thanks. We'll be right there," said Jeremy, trying to control his smile. They got out of the car and started walking toward the church. Jeremy put his arm around Jason's shoulder. They were practically the same height. Jason's arm soon followed. If "I don't know" wasn't the answer he had hoped for, no one would have known as they walked toward the church.

They stood outside, both in their sunglasses. The Bishop boys. Jeremy again said all the niceties to the strangers who knew him. He and Jason walked into the church. It was evident that the "in lieu of flowers, please donate to the Bishop Scholarship Fund" had been ignored, for the casket was surrounded with wreaths. As they walked to the front right pew reserved for the pallbearers, Jeremy focused on the journey of Christ depicted in the illustrations hanging on the walls between each window.

It had been planned that the service would be as brief as possible. Two of his father's friends spoke. The choir sang. The reverend preached, on "Set Thy House in Order." Jeremy's arm perched on the back of the pew, but his hand was on Jason's shoulder. He looked over at Carol and Jessica, neither shedding any tears. They sat there, mother and daughter, their hands clutched together. They were in the numbness of denial where grief begins. Carol looked over at Jason, who cried freely. She smiled at Jeremy, pleased to see him consoling Jason.

Jeremy had sat in this very spot for Mama B's funeral. He

didn't cry then and didn't cry now. Yes, the loss was different this time, but this time, he had gained some understanding. His father was free now, and to some extent, so was he.

He listened as people spoke about how wonderful his father was. He listened as the preacher told those present that this was a joyous day, "as brother Bishop finds himself in sweet repose, free of sin." He listened as the amens fluttered as freely as clothes on a line in spring. He listened harder when the preacher spoke of this man's undying love for his family and how that love would get them through this difficult time.

When it was over, friends and family went up to pick up a wreath and carry it down the center aisle to the waiting hearse. Paul picked up one with carnations, and as he passed Jeremy, he whispered, "Wassup?" When all the flowers had disappeared, just the casket remained. No distractions. No words. Once again, there was nothing between him and it. It was then time for Jason's arm to be his support.

The funeral attendants began rolling the casket down the aisle. Jeremy's fingertips slid along its side as it rolled by. Even after it had completely passed him, his arm didn't fall; it just remained extended there in midair, as though his fingers could still feel the surface. Oh, yes, the cold had passed, but the cough lingered.

It was Carol who touched his hand, and he pressed firmly on hers and then released it. The blood began to circulate once again. He turned to the back of the church and the casket was in the vestibule. The light from outside wrapped it in a golden aura. He stood up and pulled on the white gloves that the funeral director had given him. Jason stood beside him, doing the same. When their gloves were on, they together walked toward the light.

The police cars turned on their blue flashing lights. Rory was driving one of them. He waved at Jeremy, and Jeremy mirrored the motion. Jason turned on the Jaguar's headlights, then started the car, pulling behind the limousine. The funeral procession began, car after car. Other cars whose drivers were just going about their day pulled over, showing their drivers' respect for whoever it was being tucked away. Jeremy appreciated this gesture more than ever, for traffic stopped for no one in New York.

When the convoy made a right turn, it allowed him to see the multitude of cars. He could see their lit eyes as far as his own could see, and it amazed him that this man, his father, had touched so many people, and that they came to give their "final farewell."

"I do think of you as a brother."

"You don't have to say that."

"I know. And I love you."

"Me, too," said Jason.

That was enough for Jeremy to hear. If nothing else, his father had brought him and his brother together, allowing him to say what his father had never said to him, nor he to his father.

The cemetery was beautiful, if indeed you could put out of mind that that was what it was. Elysian Gardens. It was a lawn that his father would appreciate. Jeremy didn't look for markers this time. He was certain he would remember.

They got out of the car and walked toward the hearse. Jeremy saw the man who was his grandfather. He looked as he had the night before. Same woe-filled suit.

"I'll be right back," said Jeremy. He walked over to the man, who was keeping his distance. Jess watched Jeremy, and Gladys again turned up her nose at something foul.

"Hello," said Jeremy, extending his hand, "I'm Jeremy."

"I know," said the man. "I'm not here to make no trouble. I just wanted to—"

"That's fine. I'm glad you're here." A pause cushioned them, then Jeremy said, "I'm sorry, but I don't know your name."

"Anthony. Anthony Jeremiah."

"You don't have to call me Jeremiah. I go by Jeremy."

"No," the man said, "that's my name. Anthony Jeremiah."

"Oh, I see," said Jeremy, finally letting go of the man's hand.

He looked at the man's face, searching for traces of himself. He saw someone who had lived a life wondering and wandering, but that appeared to be the only similarity. Jeremy took the man's arm just below his elbow and led him to where the other people were assembled. He smiled at Jess; she smiled back.

He went back over to Jason. They put back on their white gloves. The casket was rolled out. Jeremy took the first rung on one side, while Jason had the other. They led the walk to the grave. Once there, they placed the casket on the straps over the opened earth. A green cloth covered the dirt that would later fill the hole. Jason went to stand behind his mother, and Jeremy, behind Jess.

The reverend spoke of man going back from whence he came. In less than five minutes, it was over. Jeremy took the red rose from one of the funeral attendants and asked for another. The crowd began to disperse. Jeremy sat in one of the family chairs, oblivious of everyone or thing around him. He wanted to see the body lowered. He didn't know why that was so important to him, but he wanted to see the casket go into the earth, to make certain that it was handled with care.

"When do you lower the body?"

"Well, Mr. Bishop, we don't do that until everyone leaves."

"Why?"

"Well, in our experience, it's just easier that way."

"For whom?"

Jeremy didn't wait for an answer. He got up from the chair and placed the rose on the casket. He turned around and walked over to the man who was his grandfather. He didn't say anything; he didn't want to make promises or know any more. He just gave the man the other rose, then walked away.

"Ol' man was heavy as hell," said Jason, turning up the air conditioner in the car.

"You're right. I was really surprised."

"I felt my hand slippin' and shit. All I could think was *Please, God, don't let me drop this casket.*"

The funeral was over. They had made it through.

The reception was being held at the Rotary Club's banquet hall. When it was well underway, Jeremy walked over to his brother.

"Jason, Paul's gonna give me a ride to your place. Can I borrow your car?"

"You can take the Jag; I'll get a ride back."

"No, that's all right. I rather take your car."

Jason didn't ask any questions. He pulled out his keys and took off the one for his car. He handed it to Jeremy.

"Thanks."

chapter 19

⤳ ⤶

When the old blue BMW pulled into the cemetery, it was as though it knew where it was going. Jeremy didn't count the trees; he just drove. He parked the car, then walked toward the grave, thinking how quickly they become just that.

The dirt, uncovered by the green cloth, had found its rightful domain, making a mound where the casket once was. The sun had had its way with the varied flora that had been relieved of their cards so the appropriate thank-you notes could be written. He had been amazed by the number of roses and lilies and the "God-awful carnations" that had been sent. So many people were familiar with this man, yet Jeremy had not so much as uttered a vowel or consonant in favor of him. But standing there under a gravity-less moon, the sun heading to bed, his time had come. Sunset had always been his favorite time of day, for it presented the best light.

He sat on the ground next to the grave, somewhat amused by the melodrama of it all. He saw a man walking toward him, carrying flowers.

"Howdy," said the man, the flowers at this side.

"Hi."

"You knew Mr. B?"

"Yeah."

"Good man. I'm shole gonna miss him a helluva lot. Salt of the earth, that one there. Told me he was sick, but he'd made 'rangements for all the plots. Paid me a little som'n extra to see that they were taken care of and fresh flowers were put out. Said I could put anythin' on 'em but carnations. Man hated himself some carnations." Jeremy couldn't even laugh. He wanted to cry, but that wouldn't come either. "I was sad to hear the news. Ill health just don't care 'bout age. But he loved his wife som'n special. It'll be the first time in a long time that I'll have to look at my watch to see it's eight."

"Eight?"

"Oh yeah. He come ever Sunday. Eight o'clock on the dot. You could set your watch to it. I know all the families 'cuz I'm out here ever'day. Most of'm come once, twice a year. But Mr. B'd stand out here for an hour or so, sometimes longer, then come and sit with me in the little shack for a while. Rain or shine. I'ma miss him somethin ter'ble, but I know he would have wanted me to bring these out here today rather than tomorrow."

The man put the flowers not on his father's grave, but on the plot next to it. He said his good-bye, then walked away. Jeremy looked at the plot. It had been concealed earlier by the flowers and the people and he hadn't noticed. A cold sweat broke out over his body as he stood up to read the headstone.

No thoughts filled his head, and he didn't notice the woman walk up behind him. She wore a black dress with a single strand of pearls, and a black hat with veil covered her face. She carried red roses pressed against her stomach. The image seemed like a photograph.

"Hello," she said.

"Hello."

"I didn't expect anyone to be here now."

"Nor did I."

"You look just like your mother."

"Thank you," he said, as if he'd heard it often. "I didn't realize that she was buried here."

"A wife is always buried next to her husband." She placed the flowers on his father's grave. "This was your mother's. I'm sure she would want you to have it." The woman handed Jeremy the rosary she had been holding and walked away, leaving him as she found him, but changed as well.

He didn't try stopping her. She got back into the car and the driver closed the door. He watched as the car circled the innards of the cemetery, then drove away. He held the beads in his hands and felt each one, and the only prayer that came out of his mouth when he flung his head back was "Jesus Christ," in a sigh that wafted into the night.

"I've gotta turn on the sprinklers," shouted the man from the little building.

"Just a few more minutes, please."

"Just take your time," the man said, then added, "You're his oldest, ain't you?"

"Yessir."

"Jeremy, right?"

"Yes."

"Always talked about you. Some proud, he was. You're lucky to have a father that loved you as much as he did."

"I know."

Their exchange was vollied over many marble-chiseled lives and memories, for every one of those souls to hear. Jeremy didn't cry. No, he wept. For the first time, he did believe that his father loved him, and it seemed only appropriate that once again, a stranger had impressed it upon him.

"I love you," he said, staring at the mound of dirt. "I never told you and you never told me. I'm sorry about that. But I'm saying it now, and I know you hear me. Too many years, too much time. But that said, I can't so easily forget. I'm not man enough to forget just yet. One day, maybe, but not right now."

He reached down and grabbed a handful of dirt. As he let it escape his fingers, the wind's invisible breath blew the dust onto him, but not into his eyes.

As he walked back to the car, he could see the sprinklers coming on in the distance. The faint hissing sound they made reminded him of the release, and he had no more tears. The loss of innocence no longer bothered him.

Jeremy was driving down Highway 165. He felt more relaxed than he had in some time. He turned on the radio in the BMW, moving the dial that presented various sounds. One station surprised him with "Rhinestone Cowboy," and the words poured out of his mouth like old times. He cranked up the volume, and his voice was so loud that he didn't hear the horn of the car behind. The car sped up and pulled next to him, and an arm was waving out of the window. Out of the corners of his eyes, he saw it was Gloria. She motioned for him to pull over, and he did.

"Hey, baby," she said, walking toward his car. "Lawd, have mercy. Look at you."

"No, look at you," he said. Gloria had her hair closely cut in a natural. She was still pleasingly plump but wore a dress rather than polyester pants.

"What can I say? Milk does a body good," she said, spinning around for him.

"Missed you at the wake and the funeral."

"Well, I know, and I'm sorry, but I was kind of busy," she said, turning back to her car and motioning for the man to get out. He walked over to Jeremy. "Jeremy, this is Maurice. Maurice, say hey to Jeremy."

"Hey."

"Awright. Go back and wait in the car."

Maurice went back to the car.

"Well, I see you have him well trained."

"Wouldn't have it any other way. *Bam!*" she said, putting her hand up in Jeremy's face to reveal a ring on her third finger. "He kept on tellin' me it was time for us to settle down, and I kept tellin' him, 'Ain't no rings on these fingers,' and *Bam!*" She again presented her hand for his viewing.

"Don't tell me you're married."

"Nope. Engaged. He did it last night. Got down on his knees and all. Now, he'd been down there before, but not with a ring, if you get my meanin'."

They laughed and hugged as though this was as normal a place to do so as any and the cars passing didn't exist.

"I did want to come and pay my respects to ya daddy, but I just didn't think it would be right to go to a funeral the day after ya get engaged. Happy people ain't got no place at a funeral."

"I suppose you're right. Have you told Aunt Jess yet?"

"No. I haven't told nobody, which ain't like me, but I'm keepin' my mouth shut until after I have the ring checked out. It looks real, but ya can't be too sure." Gloria turned back to the car and waved at Maurice, and he waved back. "Just as sweet as he wanna be."

"Well, congratulations."

"Now, I know I ain't as young as I used to be, but I've finally found somebody that can put up with me. He don't say too much, but he makes me smile and I love him and he loves me. It's a good feelin'."

"Good for you. You deserve it."

"Yes, I do," she whispered, as though saying it to herself. She was no longer hiding behind "lemmetellyasom'n."

"And ya know ya gonna have to come back down here and take pictures at our weddin'."

"Well, Gloria, I really don't do . . . Sure. You just let me know, and I'll be there."

"Ya promise?"

"I sure do."

Gloria began to tear up. She looked at him as though he too had proposed to her. They hugged, and she slid her hands down his arms until they were holding hands. She then slid hers away, then walked back to her car. She closed the door and kissed Maurice on the cheek. Their smiles were as bright as day. He stood there as they pulled their car away, passing him. Gloria had her hand up, waving backwards, again displaying her ring. Jeremy gave her a thumbs-up then got back in the car, and his words to her rang through his mind.

You deserve it.

"Yes, I do," he said, pulling the BMW back onto the highway.

"I was worried about you," said Carol as Jeremy sat down with her at the kitchen table.

"I'm actually all right."

"I'm not at all surprised."

"How are you?"

"We'll see. I'm going to attempt to sleep in the bed tonight. That'll be the first step. Then I'll go from there."

"The first step is always the hardest, isn't it?"

"Always."

"You mind if I crash here tonight?"

"Of course not. This is your home. You never need ask."

"Thanks, Carol. Are Jason and Jessica back there?"

"Yes. I think they're in their rooms."

Jeremy walked to the back of the house. He knocked on Jessica's door.

"Come in," she said.

"Hey."

"Hey, where've you been? We were worried."

"I went back to the cemetery."

"You are an odd bird."

"You're just figuring that one out?"

"Hey, man," said Jason, walking through their bathroom door. "I thought I heard you in here. I hope you did the big brotha thang and hooked me up with a tank of gas."

"Can't you knock?" snapped Jessica.

"Shut up, Thelma."

"No, you shut up, ya punk."

Jeremy started laughing, and they stopped arguing and looked at him.

"What do you think you're laughing at?" asked Jessica.

"The two of you. You make me laugh."

"Oh, yeah," said Jessica, picking up a pillow and hitting him. Jason joined in and a pillow fight began, the first one Jeremy had had in his twenty-six years. He fell on the bed, and when they had each exhausted all the breath they had, they all lay there together.

"So when are you all coming up for a visit?"

"When we get an invitation," said Jason.

"Well, you've got it."

Jeremy lay with them for a while, and it was like he had known them all their lives.

"Are you two gonna be all right?"

"No sweat," said Jason.

"Jessica?"

"It'll take some time, but that's what it takes."

"Thank you, Black Abby," said Jason, and the pillows again started swinging.

Carol came to the door to see what the noise was about, fearing it, yet seeing that it was without anger, she too took a pillow and joined in. Her blows were the strongest.

Paul knocked on the garage door. At last Jeremy felt as though he was at home, so he went to answer it.

"Hey," said Jeremy, stepping outside.

"Thought I'd come see if you were still here."

"Yeah. I wanted to hang with Jason and Jessica for a while."

"So you outtie in the mornin', huh?" said Paul.

"Yeah, back to the hustle and bustle."

"Back to champagne wishes and caviar dreams and all the women—and probably a few men—after you."

"Well, you know, every job has its perks. It's all about flexibility."

"Yeah, I bet it is," said Paul with a wink. "But you gonna have to commit at some point. I don't know if you believe it, but somebody's gonna be damn glad to have you."

"I hope so."

A moment passed, leaving Jeremy with that hope, then Paul said, "You went back out there, didn't you?"

"How'd you know?"

"I did the same thing. I left a lot of hatred buried in the cemetery. It was like Pat and I were going at it for the last time and there would be no winner."

"I saw my mother's grave," said Jeremy. "It had my birthday carved right on it. She'd been out there all this time and no one knew. He went out there every Sunday. Carol thought he was seeing another woman. Then, if that doesn't beat all, my grandmother, or at least my mother's mom, came to put flowers on his grave. It was surreal. I didn't ask her anything. I didn't know her enough to be upset."

"Oh—here," said Paul, pulling out a pack of Camels. "A l'il goin'-away present?"

"Thanks, but I think it's time for me to stop. Those things'll kill ya."

"Yeah, I guess it's time to move on. You need a lift back over to Jess's?"

"Thanks, man, but think I'm gonna hang here tonight."

"Well, I'm not gonna come out to the airport tomorrow. I think it would be better if it was just *la familia*."

"You're family."

"Thanks, man, but you know what I mean."

"Yeah. And listen, don't be a stranger."

"You neither. I know Elsewhere ain't as hip as the Big Apple, but you could come down here too, you know."

"I will, but planes fly both ways."

They hugged, and Jeremy walked Paul to his car. Paul started it up and backed it out. Hip-hop music blared from the window, then he turned down the volume.

"Hey!" screamed Paul from the car, his finger pointing to the sky.

"What?"

"Don't forget to look up."

Jeremy waved and watched as the car disappeared. He stood viewing the night. He found the Little Dipper and the Big Dipper and all the other named nameless stars, and he did appreciate them.

"I won't," he said to the night.

I called your Aunt Jess to tell her you were spending the night."

"Thanks, Carol. How was she? Do you think I should go over?"

"No, I think she's fine. Your Aunt Gladys left a couple of hours ago, so I'm sure she's probably dancing a jig."

"Carol," said Jeremy, stopping the sentence as though trying to find the words, "every Sunday, Dad went—"

"I know," she said, stopping him. "A friend and I followed him one morning. I had suspected he had been seeing another woman. I was right."

"Why don't you think he ever told you?"

"Your father was a loving man, with a lot of pain. I know you get tired of hearing that, but it was true."

"I know."

"How could you?" she asked. "You never got to feel it. I did, and when you feel it, you know it's real. I felt his love for you, and that's why he didn't tell me he was going out there. He had rather lie to me than make me feel I was second best. I never thought I could replace your mother in his heart or in yours, and I never tried to. But he had a big heart with plenty of room. He let me in, and it wasn't easy for him to do so. He was a proud man. He didn't want anyone to see him suffer. He didn't want you to see him suffer. He's been a great father to the kids. I know you probably don't want to hear that."

"No, I'm glad. They're great. And I did all right. I had Mama B and Aunt Jess."

"You sure did. I can see why she called you Patience," said Carol, getting up from the sofa. "It's been a long day. What time is your flight?"

"Ten-forty."

"Okay. I'll see you in the morning."

"Good night, Carol. Sleep well."

Jason and Jessica rode to the airport in the Oldsmobile with Jeremy, and Carol and Jess followed in the minivan.

"See you soon, ol' girl?" he mumbled as he passed where the third tree used to be.

"What?" asked Jessica.

"Nothing. Just thinking out loud."

He checked in and dropped off the keys for the rental car. He was going home.

"Now, don't you be a stranger, hear?" said Jess.

"I won't, but I still want you to come up."

"I'll think on it."

"Jason, I want you to take care of these lovely ladies."

"I got it covered."

"He needs to try taking care of himself," said Jessica.

"Shut up."

"You shut up."

Jeremy looked at Carol. He thought how lucky his father had been to have her, and he knew he couldn't share the same pain she felt, but he could assist in trying to ease it. She hugged him and whispered in his ear, "One step at a time." They broke their embrace. Jeremy walked a few steps, then turned back and looked at them. They all stood smiling. For the first time on this trip, he wished he had his camera—and the thought of color came to mind.

When he landed at La Guardia, he looked for his name placard, indicating his driver. He stopped when he saw that it was the same driver who began the journey with him. "So, three get-it girls . . ." began Jeremy.

That afternoon when I walked into my apartment, it looked different, more than just a filled empty space. I walked past the answering machine and its blinking light, not quite ready for what that might entail. I needed a moment to myself. I just wanted to sit in my rocker. I swayed back and forth and when my pulse eased to the flow, a knock came at the door.

"Here's your mail, Mr. B," said Jimmy. "And this box arrived this morning." I threw the mail on the desk and looked at the box. I didn't hear Jimmy after that. He just closed the door behind him, letting himself out. I placed the box on the table, then took the letter opener from my desk drawer. I sliced the tape that kept the four flaps in place. When I opened them, there was my high-school diploma and dozens of magazines with page markers. Each one revealed a photo that I had taken, even some that I had forgotten about.

Under the magazines were pictures of me, probably every school picture that I'd ever taken. I looked at them, remembering each year and the reason I hadn't smiled. When I got to the bottom of the box, there was a picture frame. I saw the back of it, and my heart no longer pumped to my brain, but became it. I didn't want to pick up the frame, yet I did. I held it for a long while, walking around the room, refusing to look. Finally, I slowly turned it over. It wasn't the picture that I had hoped for. It was my birth certificate. I really looked at it. It was the first time that I'd ever seen the three names together. Ours.

Helen Wilson Nichols.

Christopher Van Bishop.

Jeremiah.

It was the closest thing to a family photo that I would ever have, and a face no longer seemed important.

I walked over to the mantel and set it there. I looked at the mirror above it with both eyes opened and for the first time, the reflection that I saw was all that I needed to see. I felt centered and I knew that was because it's difficult to maintain any sense of balance with your eyes closed.

I did finally listen to the messages. The usual suspects. But the last one was from Paul. Beth had delivered. "All ten fingers and toes and the eleventh one." As I was about to pick up the phone to call him, my hand was met with a ring. I let the answering machine pick it up. I just wasn't ready to jump right back into the fray. Carol's voice came on the machine.

"Carol? Hey," I said, picking up the phone. "Sorry, I was screening."

"That's fine. Did you get the box?"

"Yes, I did. Thanks. I can't believe he had all those things."

"Pictures can never replace the real thing. But when you feel it's all you have, it has to be enough," said Carol, her voice quivering. "I'm going to get off now before I start crying."

"Are you okay?"

"I'll be fine."

"Call anytime, all right?"

"Thank you, Jeremy. Bye-bye."

"And Carol," I said, after hesitating—but, as is often the case, the hesitation was too long and the dial tone filled my ear. Yet as my arm came down, I found myself saying, "Call me Patience."

Walking Through Mirrors

Brian Keith Jackson

ABOUT THIS GUIDE

The suggested questions are intended to help your
reading group find new and interesting angles and
topics for discussion for Brian Keith Jackson's
Walking Through Mirrors. We hope that these ideas
will enrich your discussion and increase your
enjoyment of the book.

Many fine books from Washington Square Press
include Reading Group Guides. For a complete
listing, or to read the Guides on-line, visit
http://www.simonsays.com/reading/guides

DISCUSSION QUESTIONS

1. Why did Jeremy's Grandmother call him Patience? Did it fit him as an adult, or was it ironic?

2. The narrator tells how his hometown 'Elsewhere' got its name. How is the naming of the town significant to the story?

3. Jeremy's father acted distant towards Jeremy when he was a child. Why was that? How does the truth that Jeremy discovers as an adult differ from his childhood perception of his father?

4. Aunt Jess told Jeremy that she wasn't surprised that he became a photographer because he was always "looking at the camera like he was trying to figure it out." Why did he become a photographer?

5. How are his memories like photographs?

6. Why did Jeremy fear having his picture taken?

7. What is the significance of the swatches of fabric that Mama B added to the cushion of the rocking chair?

8. When describing New York City, the narrator states, "It's a great place to live — or at least work. But Elsewhere is what got me there. It prepared me. It's a love - hate relationship in its most realized form." Is there a place, or a time in your life that prepared you for your present situation?

9. What role did Charles play in Jeremy's life? What did his death illuminate?

10. Jeremy's family was Baptist, so why did he attend a Catholic church? Was this part of a bigger identity crisis?

11. Why had he not been told the circumstances of his mother's death before?

12. Why was Carol so concerned about how her husband treated his first son? How does the narrator feel about Carol?

13. What is his relationship with his siblings like? Did loss unite them, or was there another precedent?

14. Why is the novel titled *Walking through Mirrors*? What was Jeremy's father trying to tell him?

15. How will Jeremy's life be changed after the funeral?

AN INTERVIEW WITH BRIAN KEITH JACKSON

Q: Why is the main character a photographer? Do you have any background in that field?

A: In most of my work I attempt to deal with field and perception, which correlates well with photography. We had a darkroom in my childhood home and my father was quite handy with a camera, so I've always had an interest in that art form. Photography is another aspect of story telling. Yes, it may be worth a thousand words, but I take or look at a photo and try to get at least 70,000 words out of it.

Q: The narrator's sexual preference remains ambiguous throughout the novel. Why is that?

A: I wanted to focus on gender roles and issues rather than preference. With *The View from Here*, I was able to write about a pregnant woman. Of course, I will never know what it is like to be pregnant, but I try to make my work more about the recognition of human emotions that we all share. In that, I also made a conscious effort to make certain that the secondary characters were ambiguous in regard to race. We are often so emphatic about exteriors that it keeps us from acknowledging the person.

Q: You live in New York City yet write about the South, where you grew up. Do you find that separation, or semi-exile, necessary to write your novels?

A: Well, you can take the boy out of the South, but.... I have an intense long distance love affair with the South, yet I'm still working on my relationship with New York. I'm sure we'll see more from her [New York] in the future. I attempt to make locale one of the characters, and the South has a long history of character. If you can so easily escape where it is you come from, then it's difficult to appreciate where you are.

Q: Why did you choose the father's death as a catalyst for this novel?

A: In *The View from Here* I used a birth as the catalyst, so in *Walking Through Mirrors* it happened to be a death. In a sense, I see them as similar for they have a huge impact on what will or has occurred between the two. I wanted to show that issues should be resolved while people are alive so that there can be some sort of release, yet not of regret. We all have our limitations, but it is never too late to begin that quest for understanding, for renewal.

Q: What are you working on now?

A: I'm working on several projects at the moment—all of which are in the early stages, so I dare not mention them. I'm rather superstitious about that. I guess you can say that I'm courting new characters right now, waiting for their cooperation. But as my grandmother used to tell me, "Never announce the wedding until both parties say yes."

BRIAN KEITH JACKSON
has received fellowships from Art Matters, the
Jerome Foundation, and the Millay Colony for the
Arts. His first novel, *The View From Here,* won the
American Library Association Literary Award for
First Fiction from the Black Caucus of America.
He lives in New York City.